Hybrid Urbanism

Hybrid Urbanism

On the Identity Discourse and the Built Environment

Edited by Nezar AlSayyad

PRAEGER

Westport, Connecticut
London

Library of Congress Cataloging-in-Publication Data

Hybrid urbanism : on the identity discourse and the built environment / edited by Nezar AlSayyad.
p. cm.
Includes bibliographical references and index.
ISBN 0-275-96612-7 (alk. paper)
1. Cities and towns. 2. Group identity. I. AlSayyad, Nezar.
HT119.H93 2001
307.76—dc21 00-044131

British Library Cataloguing in Publication Data is available.

Library of Congress Catalog Card Number: 00-044131
ISBN: 0-275-96612-7

First published in 2001

Praeger Publishers, 88 Post Road West, Westport, CT 06881
An imprint of Greenwood Publishing Group, Inc.
www.praeger.com

Printed in the United States of America

The paper used in this book complies with the Permanent Paper Standard issued by the National Information Standards Organization (Z39.48-1984).

10 9 8 7 6 5 4 3 2 1

Every reasonable effort has been made to trace the owners of copyright materials in this book, but in some instances this has proven impossible. The editor and publisher will be glad to receive information leading to more complete acknowledgments in subsequent printings of the book and in the meantime extend their apologies for any omissions.

To the peoples of hybrid persuasions.

CONTENTS

PREFACE

The idea behind this book goes back to 1996 and a discussion I had with two of my Ph.D. students. The students were concerned that, as prospective urban historians working on Third World topics, they were unlikely to get the jobs they desired in the United States. They were also concerned that working on cities and places that were "hybrid" and principally constituted of people of different ethnicities, cultures and religions was going to further limit their chances. I was especially puzzled that two relatively bright students could feel this sense of insecurity, and I could only respond by reminding them that the choice and pursuit of a research topic has always been about the politics of position. I recognized that, like the environments and the peoples they elected to study, these students were struggling with their scholarly identities. It was as if they wanted to have their cake and eat it too. They desired the respect and security they thought would be afforded scholars working on First World environments within the confines of such established disciplines as architectural history or material-culture studies, and at the same time, they wanted the attention that would come from being creative intellectuals taking on new topics and pursuing them in innovative ways in unconventional sites. Simultaneously, they wanted "sameness" and "difference." It was this encounter that prompted me to think about the subject of hybrid urbanism, for I realized that what these students were seeking was precisely what the people who lived in the hybrid places they studied seem to have achieved.

Hybridity is a very complex notion. At its most basic level, it has meant the interbreeding or mixing of different peoples, cultures and societies. In the last few years many good books have been authored and edited on hybridity and the concept of a "third space." However, few of these have attempted to ground themselves in the concrete realities of the physical environment: in neighborhoods, housing projects, urban squares, and city streets. This is the added dimension this volume hopes to provide. Anchoring their work in specific case studies that mainly concern nineteenth and twentieth century places, the contributors to this volume present a rich cross-cultural understanding of how the notion of hybridity has contributed to the shaping of a unique urbanism, an urbanism that I prefer to call the "third place."

No book with "hybridity" in its title can ever claim to be comprehensive, representative or even consistent — qualities that all run counter to the notion itself. Because of this ideological position, I have intentionally refrained from agreeing with the contributors to this volume on any specific definition of terms. Instead, I have allowed them freedom in structuring their arguments and in forming their narratives. The process by which the book was put together was also simple, consisting principally of compiling pieces which were products of my personal intersections with the contributors. One-half of the contributing authors are current or former students whose chapters were originally written as papers for my graduate seminars or as revisions of portions of their dissertations; the other half are colleagues who participated in various conferences and symposia which, in most cases, I have helped organize during the last four years. Their contributions generally consist of revised versions of papers they initially presented at those events.

Many people deserve my sincere appreciation for helping put this book together. First, I would like to thank the contributing authors for mostly abiding by my deadlines and requirements, and for the leverage they afforded me in editing their work. Duanfang Lu, my principal research assistant on the project, deserves special recognition. She spent many hours, first in the library painstakingly fetching unusual sources, and then on the computer inputting and often summarizing difficult material. I have benefited intellectually from our meetings and I am thankful for her diligence. I would also like to thank David Moffat, my editorial advisor, for his effort in preparing and revising this manuscript. David has worked with me on numerous projects since 1987, and I am grateful for his patience and dedication. Annie Tennant picked up where David left off, and I am grateful for her copyediting input. Finally, Annabelle Ison, who, like David, has worked with me since the late 1980s on many publications, deserves special credit for the design of this book. I have always valued my association with Belle, and I am delighted at the outcome. Ananya Roy, who has been my principal research associate on various scholarly endeavors over the last few years, has challenged many aspects of this book and identified many problems with it. Her criticism prompted me to revise both the book's structure and my own contribution to it. I am intellectually indebted to Ananya, and I am grateful she agreed to write an eleventh-hour epilogue. At Greenwood/Praeger, Cynthia Harris handled the book in its final stage and I would like to thank her for her initial trust in the project. I would also like to thank Lisa Webber, Heidi Straight, and Diane Burke for their sensitive management of the manuscript at the press.

Finally, during the three years I have worked on this book, I have benefited from my association with several institutions on the Berkeley campus: the Center for Environmental Design Research, the Center for Middle Eastern Studies, and the International Association for the Study of Traditional Environments. I am grateful to my colleagues, staff, and graduate-student assistants in these units for their assistance. U.C. Berkeley has been, and continues to be, one of the most fertile sites to pursue such scholarship.

Nezar AlSayyad
Berkeley, August 2000

PROLOGUE

Hybrid Culture/Hybrid Urbanism: Pandora's Box of the "Third Place"

Nezar AlSayyad

> He loved the contrast between Mexico and California. An exuberant San Diego filled with impeccable courses, right next to a barren Tijuana where dust devils never gave local housekeepers a minute of rest. Scott felt comfortable in a place where the confusion of identity created hybrid town names, like Mexicali and Calexico. Where Chinese ate their chili dogs with green tea. Where someone had the sauciness to print "Juan López — Smuggler" on his business card. Where a dental technician brought up in Faribault, Minnesota, could find her Latin lover in an illegal immigrant who came from a tiny ranch somewhere in Jalisco. In Tijuana, Scott had screamed through the fanciest residential neighborhoods at three in the morning. He had picked several fights at El Reventón, always knocking someone unconscious. He had mingled from table to table, kissing other men's women. He had drunk himself beyond self-respect on countless occasions. And he had gotten away with it all.
>
> The United States wasn't the ultimate free country after all. Mexico wasn't either. Instead, it was that piece of land where one met the other in an unavoidable frontal collision. But what he really loved about the border was that he could be a respectable, law-abiding, and law-enforcing San Diego judge and, in a matter of minutes, become an unidentifiable being sleeping in the arms of a prostitute, in a place where no one cared if what he did was right or wrong. And now that he was bringing Esperanza across the line with him, the parallel sides of his life were about to merge.
>
> — Maria Amparo Escandón, *Esperanza's Box of Saints*[1]

Scott and Esperanza are characters in a popular novel written originally in Spanish by a Mexican-born American novelist, and published in the United States in the last year of the twentieth century. Describing the lives of several individuals of different original cultures whose paths intersect in the pursuit of happiness or salvation, the novel presents this time and space of intersection in terms which I believe capture the spirit of the so-called "hybrid place."

In recent cultural studies, hybridity has been discussed extensively, usually in abstract ways. But hybridity must also be understood in the context of everyday practices. In fact, I believe that it is only in this context of real life experience that it becomes possible to unravel this complex notion.

Today, in any significant urban center around the world, manifestations of hybridity are self-evident. This not only is the case in the cities of the First and Third Worlds but also in much public discourse on issues that range from industry to arts, from business to media. A few recent articles from the *New York Times* serve to illustrate this point. On May 27, 1999, Matthew Wald reported that the annual showcase for electric cars was particularly crowded because of a new emphasis on "hybrid cars." His article, published in the "Business Day" section of the paper, spoke of a new model that would surprise purists because it used batteries and electric motors, as well as gasoline or diesel engines. While other electric cars had failed to catch on because of their high cost and poor performance, Wald reported the new hybrids were "clean," "peppy," and able to go "twice as far as a conventional car on a gallon of gasoline."[2]

A month later another article in the newspaper, this time in the "Living Arts" section, demonstrated that "hybridity" was also an underlying theme in popular movies such as Disney's *Tarzan*. In his essay "From Darwinian to Disneyesque," Edward Rothstein offered a discussion of hybridity as a central feature of the Tarzan story and its evolution.[3] Rothstein explained that in the original *Tarzan* novel, written by E.R. Burroughs, the main character, Lord John Clayton, is commissioned by Her Majesty's Government to investigate a competing imperial power's "unfair treatment of black British subjects" in a West African colony. "The black natives [were] being indentured into a form of slavery," and Clayton was dispatched to "see justice done." Burroughs generated a series of events which left Clayton and his pregnant wife lost in the jungle, and whereby Clayton's first vision was "of a fearsome manlike figure in the distance, which [turned] out to be a vicious ape, the inverse image of his future son, an apelike figure who [turned] into a man."

Rothstein went on to explain how in the Disney version of the story, "the apes end up seeming more sympathetically human than their hunters." And, mindful of the Disney mythology, he argued that almost every animated Disney film has "important ethnic characters" and honors a version of "melting-pot liberalism" in which "[t]he outsider is welcomed, the tramp civilized, the beast turned human, the mainstream transformed." Rothstein observed, however, that this theme has recently changed under the impact of ethnic politics. Instead of seeking integration, films are increasingly preoccupied with identity and separation. For Rothstein, Tarzan seemed a kind of hybrid of earlier Disney modes, "an outsider who yearns to be accepted in a society he seems ill suited for."

> He is heartbroken over his difference from his jungle family, but firmly declares, "I'll be the best ape ever," an immigrant determined to earn his credentials. . . . The ape king, who rejects the pale-skinned human boy, is challenged by Tarzan: "Why are you threatened by anyone different from you?" But then comes acceptance; at the point of death, the ape king names Tarzan heir: "Take care of our family, my son." This melting-pot acceptance is then joined with a . . . celebration of a beleaguered ethnicity: taking care means defending the clan against the alien, white and human intruders.[4]

According to Rothstein, "Tarzan cannot fully oppose the human nor fully join the apes. He is like the child of a biracial or bicultural marriage. The ethnic issue is by no means clear; allegiance can be declared, but it can't be made absolute." Thus, the movie ends with the decent human deciding to remain in the jungle because in the final analysis, Tarzan is "right at home." This leads Rothstein to conclude that "the Disney myth, which originally celebrated the melting pot, now honors the latest version of democratic multiculturalism."

A third dimension of the hybrid appeared the same week in the "International" section of the same paper in a story by Joel Greenberg entitled "Combining Two Identities: Arab in Israel Parliament."[5] The story described how one Israeli lawmaker, Dr. Ahmad Tibi, stood in silence as others sang at the swearing in of members of Israel's new Parliament. Tibi was caught in the contradiction of his twin identities as an Israeli, whose state does not recognize the right of Arabs within its borders to their own state, and an Arab, whose larger political culture has repeatedly called for the destruction of the state he now serves. The article explained that although one of every six Israelis is an Arab, no Arabs have ever been named cabinet ministers, and only 5 percent of employees in the Israeli civil service are Arab. When Tibi expressed interest in serving on an important Parliamentary panel dealing with security issues, rightist members of Parliament objected, suggesting that his loyalty was suspect. Nevertheless, with the practiced skill of a man used to straddling the fence, Tibi had succeeded in his Parliamentary election by declaring "I don't have dual loyalty, and my identity is well defined. . . . I am both Palestinian and Israeli." Packaging himself as "a Palestinian Arab citizen of the State of Israel," Tibi has attempted to reverse his status as an individual from a deprived national minority with no influence on national decision making. He has announced that he will become "a partner with the Jewish majority in all civic functions, which can ultimately lead to civic equality." But he has added, "I want integration, but not assimilation."

What may be observed briefly in these articles are situations in which assumed dualities are torn open and the logic of hybridity is introduced. The process does not simply involve the combination or merger of incompatible elements, but instead the insertion of a third possibility connecting originally incommensurable terms and irreconcilable realities. The cases of the hybrid car, Tarzan, and Ahmad Tibi have certain things in common: in all three the subjects are assigned "identities" that defy the norm; their existence presents a problem to the dominant majority, be it an industry, category or a group; and their perceived difference offers them the opportunity to challenge both the hegemony of the norm and the power of the dominant majority.

It is striking that during the few months in which these articles appeared, "hybrid" stories were told again and again, in one section or another of the *New York Times*. It seemed that the discourse on hybridity had established itself as a common subject. What are the implications of this normalization of hybridity, and consequently its legitimation in the sphere of urban space? Is it a welcome tendency toward the globalization of the exclusive notion of identity? Or is it simply a new political device that aims to unsettle traditionally established political and social norms?

IDENTITY AND HYBRIDITY

To be able to answer the above questions, I start with the more familiar theme of identity. The contemporary meaning of the word developed from diverse origins. The *Oxford English Dictionary* explains that "identity" derives from the Latin *idem*, which means "sameness," and that it shows an etymological similarity to "likeness" and "oneness," which are expressed by the Latin *similitas* and *unitas*. Likewise, "ident(i)" stems from the Latin *identidem*, meaning "over and over again, repeatedly."[6]

As a subject, identity at the end of the twentieth century is often addressed as problematic. For example, Kathryn Woodward has suggested there is a "crisis of identity" that has emerged from the breakdown of previously stable group membership. Identity now increasingly derives "from a multiplicity of sources — from nationality, ethnicity, social class, community, gender, sexuality — sources which may conflict."[7] Thus, despite the fact that "identity" may be rooted in "sameness," Woodward has claimed that identity is always about difference. Stuart Hall has presented a similar position in his essay "Who Needs Identity?" Drawing on the Derridaian notion of *différence*, he has argued that identification as a process always operates across difference: "[I]t requires what is left outside, its constitutive outside, to consolidate the process." Hall has thus argued that identity should be understood as "produced in specific historical and institutional sites within specific discursive formations and practices, by specific enunciative strategies."[8]

Although there may be no fixed notions of identity, all involved in a political culture will behave and respond as if there are well-defined, historically established ones. Thus, as Benedict Anderson has successfully demonstrated, all national identities are constructed, and the differences between them lie mainly in the different ways in which they are imagined.[9] At a practical level, one might further note the circular nature of the relational logic of identity. By this I mean that discourse on identity will always take people to the "other" only to bring them back to themselves. In this discursive process, the framework of reference shifts continuously, but it always returns to the self. It is "I" versus "you," "you" versus "them," and then possibly "them" versus "us." However, for the purposes of this chapter, I discuss this complex notion of identity only as it relates to three other notions: hybridity, difference and place.

I am engaging here in an exercise of keywording following Raymond Williams.[10] The meanings of keywords are often inextricably bound up with the problems or issues they are used to explain. As such, they may be used to represent an entire vocabulary of shared meanings. Nevertheless, I believe etymological consciousness of the words themselves is essential to understanding the social and intellectual context within which they are used.[11] According to the *Oxford English Dictionary*, "hybrid" originally derived from the Latin for offspring "of a tame sow and wild boar," or "of human parents of different races." However, today its meaning is far broader, embodying "[a]nything derived from heterogeneous sources, or composed of different or incongruous elements." Thus, in philology, a hybrid is a composite word formed of elements belonging to different languages. In geology, its use was defined in the 1950s to describe "the products of intermingling of . . . magmas," including all rocks resulting from "the assimilation or melting of solid igneous rocks by later intrusions from the same source." And in biology, the hybrid is "[p]roduced by the inter-breeding of two different species or varieties of animals or plants."[12]

Of the other concepts I am concerned with, let me begin with "identity." The concept typically refers to social labels given to individuals as members of a group, which may either be assigned by the individuals themselves or by others. Such a view implicitly recognizes that each person may live with a multiple sense of self and may carry disparate and contradictory identity positions. This tendency renders the "constructedness" of identity more important than its "rootedness."[13] If identity seems to resist hybridization and constitute oneness out of inconsistent strands, hybridity tends to juxtapose and fuse objects from different and normally separated sources. On the other hand, if one stresses the rootless and floating side of identity, hybridity and identity become less incongruent. Thus, to paraphrase Steve Pile, hybridity and identity, "whether considered real or not, do not arrive out of nowhere, they are socially constructed from the struggle to maintain this dissimilarity."[14] To this I would add, based on some of the arguments in this volume, that whereas identity may respond to the temporal pressure for uniformity, hybridity can manifest the spatial dimension of multiplicity.

Historically, the use of the word "hybrid" has been deeply embedded within scientific racism. As Nikos Papastergiadis has pointed out, the historical association of hybridity "bears the dubious traces of colonial and white supremacist ideologies."[15]

> [A]s long as the concepts of purity and exclusivity have been central to a racialised theory of identity, hybridity has, in one way or another, served as a threat to the fullness of selfhood. The hybrid has often been positioned within or beside modern theories of human origin and social development, mostly appearing as the moral marker of contamination, failure, or regression.[16]

Robert Young has provided a detailed account of this issue in the opening chapter of his *Colonial Desire*. According to Young, the concept of hybridity was first raised in the eighteenth century, when "the different varieties of human beings were classed as part of the animal kingdom according to the hierarchical scale of the Great Chain of Being."[17] Within this classification system, the African was placed at the bottom of the human family. The prevailing view at the time was that the very idea that humans could be of different species conflicted with the biblical account. As a result, the term "hybrid" was not used to describe humans until the mid-eighteenth century. Only from the 1840s on did the belief in the possibility of human hybrids arise.[18]

Young next summarized a range of positions that have developed on the issue of human hybridity over the last 150 years. These have ranged from a species argument denying the biological possibility of racial mixing, to an amalgamation thesis sometimes accompanied by "melting-pot" notions of a new mixed race, to amalgamation's opposite, the view that miscegenation would produce only "raceless chaos," degenerate and degraded, threatening the vigor and virtue of the pure races.[19] According to Young, all these positions demonstrated that the idea of race has been "profoundly dialectical," in that "it only works when defined against potential intermixture, which also threatens to undo its calculations altogether."[20]

In this discussion, Young made a distinction between the usage of the concept of the hybrid in the nineteenth century and in the late twentieth century. Whereas in the nineteenth century the word mainly referred to a "physiological phenomenon," in the twentieth century, as the scientific basis of racism has been largely discredited, it "has been reactivated to describe a cultural one."[21] While noting the positive features of hybridity in late twentieth century discourses, Young warned that the move "from biologism and scientism to the safety of culturalism" has not been absolute, for "the racial was always cultural, the essential never unequivocal."[22] This point has indeed echoed the positions of several contemporary cultural critics on the issue. Thus, for Nikos Papastergiadis, the very idea of hybridity presupposes the ideas of an evolutionary hierarchy and biological purity, and he has argued that prior racist assumptions may now be housed in the discourse of hybridity.[23]

THIRD SPACE AND PLACE: CULTURAL DIVERSITY AND DIFFERENCE

The word "hybridity" in its postcolonial connotation first appeared in 1980s in the work of Homi Bhabha, one of the most energetic contemporary advocates of the concept. Bhabha argued that hybridity is "not a third term that resolves the tension between two cultures, or the two scenes of the book, in a dialectical play of 'recognition'. . . . [C]olonial specularity, doubly inscribed, does not produce a mirror where the self apprehends itself; it is always the split screen of the self and its doubling, the hybrid."[24] In other words, in Bhabha's Lacanian formulation, hybridity is no longer a matter of the mix of X and Y; instead, there exist X and x. In this view, x is always the partial presence of X. In Bhabha's words, "the desire of colonial discourse is a splitting of hybridity that is *less than one and double*."[25] Thus, for Bhabha, the colonizer and the colonial subject do not exist as two exclusive alternatives; a splitting of the identity positions of these two takes place through their mutual contamination. Hybridity thus does not emerge from the synthesis of different components, but from a space where elements encounter and transform each other. Bhabha believes that it is through the persistent displacements of this "in-between" space, which he names "third space," that colonial power produces its own strangers. Thus, for Bhabha, this third space had the potential to become a site of resistance, where the colonial subject hybridizes and the colonizer fails. In his later writings, Bhabha also moved beyond this notion of hybridity to address the textual construction of colonialism and the multiplicity of cultures.[26] And by grafting the psychoanalytic model onto the Derridaian formula, he has explored such familiar notions as the distinction between "cultural difference" and "cultural diversity."

In the *Oxford English Dictionary*, "difference" is defined as "[t]he condition, quality, or fact of being different, or not the same in quality or in essence." It is also defined as "the relation of non-agreement or non-identity between two or more things, disagreement."[27] In contrast, although "diversity" is defined in a manner similar to the first meaning of difference — "[t]he condition or quality of being diverse, different, or varied,"[28] it does not have the second meaning, and thus tacitly implies the possibility of unquestioned coexistence. According to Bhabha, whereas cultural diversity is "the recognition of pre-given cultural contents and customs," and so implicitly denies the possibility of hybridity, cultural difference problematizes binaries and focuses on "the problem of the ambivalence of cultural authority."[29]

For Bhabha, the endorsement of cultural diversity as the bedrock of a multicultural United States has two problems. The first is that the encouragement of cultural diversity always comes accompanied with a transparent norm given by the host society, which says that "these other cultures are fine, but we must be able to locate them within our own grid."[30] The second problem is that in societies where multiculturalism is encouraged, racism is still rampant, and this seems to be due to the fact that "the universalism that paradoxically permits diversity masks ethnocentric norms, values and interests."[31] Bhabha has therefore claimed that a liberal relativist perspective is inadequate in itself, and he has advocated a position of liminality, according to which cultural difference is no longer accommodated within a universalist framework. Instead, "[d]ifferent cultures, the difference between cultural practices, the difference in the construction of cultures within different groups, very often set up among and between themselves an *incommensurability*."[32] Here indeed lies one of Bhabha's dilemmas: he has formulated what amounts to a universal theory of the third space as a space of resistance; on the other hand, he has denied the credibility of understanding cultural difference within a universalist framework.

Katharyne Mitchell has rightfully addressed this dilemma by pointing to certain problems with Bhabha's approach to the notion of third space.

Critiquing what she termed "the hype of hybridity," she noted how this often neglects the importance of everyday practice.[33] Like Mitchell, I believe Bhabha's formulation of third space needs to be more grounded in common practices to be meaningful. Thus, one may consider Mombasa in the seventeenth century or Izmir in the late nineteenth century (or at least important neighborhoods in these cities as they are discussed in this volume) as examples of "third space." As multicultural cities before the notion was invented, they embodied the spatial reconciliation of incommensurable constructions of subcultures. Yet in contrast to Bhabha's emphasis on the transgressive nature of hybridity (that is, that it occupies a position on, or on both sides of, a boundary), I consider their hybridity to have appeared more as a condition of fundamental interaction among parties with concretely differing positions of power, who must nevertheless cohabit. As Mitchell has argued, "without context, this 'in-between' space risks becoming a mobile reactionary space, rather than a traveling site of resistance."[34]

Of course, all these issues beg the question of what constitutes hybrid urbanism. Is hybrid urbanism simply a product of hybrid culture? This is the question this book hopes to address.

GROUNDING HYBRIDITY

The contributors to *Hybrid Urbanism* have engaged the concept of hybridity in different ways. Some have taken it very literally to involve either a minority subculture that has had to accept the dictates of a dominant culture either to ensure its own survival or participate in a negotiated multiculturalism. Others have perceived it more along Mitchell's lines not only as embodying a space of ambivalent nonfixity, but also an "interlocator of resistance."[35] In such cases a prevailing "live-and-let-live" attitude among both the colonial and hybrid peoples produced the "third space" by default. Yet, as Mitchell has warned, it is important to remember that "theories that discuss space only in terms of linguistic or cultural metaphors will inevitably provide only empty theoretical 'frameworks' as well."[36] Hence, the analysis of hybrid space within the framework of colonial and postcolonial discourse must be done in tandem with analysis of specific geographic, historic and economic situations.

Anchoring their work in such analysis of specific times and places, the contributors to this volume have provided various case studies of the hybrid city which enable a richer comparative understanding of processes and events. The book is divided into two sections, with each chapter in the first section being paralleled by another chapter in the second. Part One deals with pre-twentieth century examples of places which capture the different dimensions of hybrid urbanism. Part Two takes on similar problems in similar sites in the twentieth century.

Thomas R. Gensheimer's essay "Cross-Cultural Currents: Swahili Urbanism in the Late Middle Ages" challenges the conventional view of traditional societies and their architectural forms as the product of isolated and independent developments. Gensheimer describes a world that was never isolated from cross-cultural contacts and world events even before the development of a European world system. Instead, he shows that some so-called "unique" traditional Swahili architectural and urban forms were in fact principally the result of interactions with outside influences. Thus, for example, the elaborately carved wooden doors, which are now considered a characteristic feature of traditional Swahili architecture, were derived from an architectural tradition common throughout the cities of the western Indian Ocean in the Middle Ages. The irony is that, although these doors eventually became almost extinct in the parts of the world where they originated, they remained a vibrant practice on the East African coast and came to serve as an identifying feature of Swahili tradition. Gensheimer illustrates that, more often than not, the regional cultural forms of the Swahili city arose from a hybridization process rather than a homogenization one. In other words, Swahili people did not simply adopt foreign styles and building types, but rather molded foreign architectural concepts to the established framework of indigenous institutions and building traditions in their own society to create a distinctive and unique urban form. Thus, despite certain similarities with the larger Islamic world, Swahili architectural and urban forms of the late Middle Ages represented regional developments which consistently reflected local expectations, functional needs, and spatial preferences, while still being "true" to their foreign origins.

Whereas Gensheimer emphasizes the mixing of cultures in Swahili cities, Sibel Zandi-Sayek presents the city of Izmir, in the mid-nineteenth century as a deeply contested place. In "Orchestrating Difference, Performing Identity: Urban Space and Public Rituals in Nineteenth-Century Izmir," she describes the ways in which public ceremonies and ritual events served as vehicles for building and challenging social relations. For example, the event of the Corpus Christi procession was not only inscribed in the specific sociohistorical conditions of the time, but was also framed by a variety of competing public events, ranging from Easter celebrations to the Sultan's accession. Zandi-Sayek argues that public ceremonies facilitated the transfer of allegiances from older established affiliations based on religion to a new but equally imagined form of community based on the nation-state. However, in the changing sociopolitical conditions of Izmir, there were competing loyalties: although Ottoman authorities sought to promote a unified political identity, local ethnic groups tried to advance their political and religious differences within the context of the empire. Parades and ritual events thus not only provided a medium for staging differences, but also helped draw temporary boundaries and forge convenient alliances.

Christopher L. Yip's chapter "California Chinatowns: Built Environments Expressing the Hybridized Culture of Chinese Americans" describes the development of California Chinatowns in the late nineteenth and early twentieth centuries. In particular, he describes how the social position of Chinese immigrants in the new land was an important factor leading to the creation of Chinatowns. The California Alien Land Law of 1913 banned most Chinese Americans from owning land, and realtors could not sell to the Chinese immigrants. Isolated in their ethnic enclaves, Chinese immigrants often suffered from hostility and mistreatment. California Chinatowns were thus as much a response to severe discrimination as they were to cultural difference. Yip concludes that the Chinatowns of California (and of the United States in general) were neither composed of direct importations from Asia, nor of standard building types left unchanged. Rather, they represented the transformation of existing U.S. building types to the needs of Chinese-American immigrants, and arose out of the interface between a particular group of immigrants and the new geographic settings in which they found themselves.

In "A Colonial Portrait of Jerusalem: British Architecture in Mandate-Era Palestine," Ron Fuchs and Gilbert Herbert provide a complex picture of "colonial regionalism" in a land that for centuries has been the quintessence of hybridity. Using the products of professional practice, they argue that, when British architects selected icons of Palestinian culture to inspire their designs, they did not merely comment on the nature of Palestinian identity, but took part in defining it. Fuchs and Herbert also show that, against the same colonial context, each of the architects discussed in the essay was guided by his own architectural convictions: one by Arts and Crafts ideas, another by the Beaux Arts traditions of the interwar period, and a third by his own world-embracing holistic philosophy. Thus, the work of these architects in Jerusalem resulted from the inherent process of hybridization between the British architectural theories that were part of early twentieth century European colonial thinking and the local conditions of Palestine.

The second section of the book moves to the late twentieth century, and attempts to report on the continuation of hybrid phenomena in cities. In the late twentieth century, a moment seems to have arrived when "all the old-established national industries . . . are dislodged by new-industries . . . whose products are consumed not only at home, but in every quarter of the globe. In place of old wants, . . . we find new wants, requiring for their satisfaction the products of constant lands and climes."[37] Ironically, this passage from the *Communist Manifesto* of Karl Marx and Fredric Engels originally described conditions in the nineteenth century. Thus, it would seem that nothing has changed in the intervening years. Could it be that developments in the last one hundred years have transformed the phenomenon of hybrid space from unique to typical status? The new issue today, of course, is that of late modernism and the identity discourse that has come with it. Modernist identity as an ideal type has been described as "anti-ethnic, anti-cultural, and anti-religious." As Jonathan Friedman has written:

> It works to maintain a self-experience in which alterity itself is transformed into skeptical rationalism, where nothing is taken for granted. The cosmopolitans of the turn of the century, and up to quite recently, were modernists. It was based on a striving for universal values and context freedom. The modernist nation-state was one which valued citizenship as formal membership and participation in the modern project, defined in terms of increasing standards of living, better knowledge in general, science in particular and

> high culture. As this system depended upon a hegemonic world order, which has now begun to dissolve, its modernist identity has also begun to dissolve.[38]

It is important to note that race through most of the nineteenth century was defined in terms of all-encompassing collections of cultural, behavioral and physical traits. Distinctions made between these traits were necessary to classify the "other." However, the discourse on race in the late twentieth century seems to have shifted, and Friedman has suggested it is now "itself related to the dissolution of the formerly more encompassing category."[39] Jan Pieterse has argued that the hybrid formations of the twentieth century manifest themselves in hybrid sites and spaces such as border zones, borderlands and ethnic neighborhoods. According to Pieterse, border zones are the meeting places of different organizational modes, and include such places as Free Enterprise Zones and offshore banking facilities, hybrid meeting places of state sovereignty and transnational enterprise. In recent years borderlands have become a significant topos. The ethnic melange neighborhoods of global cities have also been recognized as a form of hybrid urbanism.[40] Although the lumping together of all three categories may be problematic, it illustrates the paradigm shift that has occurred in the twentieth century, when the identity discourse has changed from that of self/other to that of historically grounded multiple subject positions. However, the problem with the postmodernist celebration and endorsement of fragmentation is that it has not fully deconstructed the spatially and temporally constructed hierarchical dualities of center and margin. "Difference" has become the new catch word, yet despite different relationships of power and histories of dominance, all groups have been recognized as equally different. The chapters in the second half of this book examine these ideas on the ground.

The second part of the book begins again in Africa. In "Stages of Globalization in the African Context: Mombasa," Ali A. Mazrui distinguishes three distinct phases in the history of Mombasa, which he calls a "global city." The Afro-Oriental phase was the period when Mombasa existed within a relatively small cultural arena, where traditions from West Asia and South Asia interacted with African traditions in search of new cultural configurations. Then, in the Afro-Occidental phase, Mombasa was "discovered" by Europeans — first the Portuguese and later the British, both of whom initiated the process of reorienting the city away from its traditional links with Asia and toward a new relationship with European culture. In the most recent Afro-global phase, Mazrui talks of a city that feels the pull of both East and West, a city that is being transformed from an "ancient" city-state to the main port of a "new nation-state." Here it becomes apparent how Gensheimer's seventeenth century Africa was the logical antecedent of Mazrui's late twentieth century Mombasa.

In "Rethinking Heritage Politics in a Global Context: A View from Istanbul," Ayfer Bartu offers an anthropological account of how the cultural heritage of Istanbul was contested in debates over the preservation and revitalization of Pera/Beyoglu during the 1980s and 1990s. Although contestations over heritage have existed during different historical periods, each group involved in the debates today accords its own meanings to heritage and to the politics of the past. The groups also disagree on which past should be preserved and to whom it should be marketed. Bartu shows that globalization has generated a new set of contingencies and possibilities in terms of what constitutes Istanbul's heritage. Rather than seeing the global/local as a binary opposition, she presents globalization as inscribed within particular localities and reworked within particular social, cultural and historical contexts. One can see how the interactions of Zandi-Sayek's nineteenth century Izmir paved the way for the debate on Beyoglu a century later.

In "Learning from Chinatown: The Search for a Modern Chinese Architectural Identity, 1911-1998," Anne-Marie Broudehoux provides an account of the century-long search for an architectural expression of modern Chinese identity. Although some scholars claim that the Chinese national style originated with the designs of Chinese architect Lu Yanzhi, Broudehoux focuses on American architect Henry Killam Murphy's attempts to adapt Chinese traditional architecture to modern needs. By showing that the so-called "Chinese national style" is indeed the invention of foreign architects, Broudehoux demonstrates that the constitution of collective identity, far from being a self-contained activity, may be truly global in nature. The question, however, remains whether a foreigner can "design" national identity. For Broudehoux, the answer is a qualified yes. The question raised by her project is whether the foreign architects who helped design a "national style" contributed to the making of a "national identity" as well. Clearly, the foreign architects who contributed to the building of California Chinatowns, as described by Yip, would claim that their work has contributed to the preservation of a unique Chinese-American identity, which is often invoked to think about the global dimension of Chinese culture.

In "Building Culture in Divided Berlin: Globalization and the Cold War," Greg Castillo presents another vision of globalization. Although the Cold War is often discussed in terms of nuclear warheads, campaigns to dominate the Third World and outer space, and the fluctuating diplomacies of belligerence and détente, Castillo employs the case study of the building of East and West Berlin to argue that culture was also conscripted by each superpower to promote its version of modern industrialized society. Whereas in East Berlin advisors from the Soviet Union equated aesthetic modernism with American imperialism, and advocated an architecture "socialist in content and national in form," in West Berlin, American advisors actively promoted a revival of Bauhaus modernism, framed as a shared symbol of progress. Most previous discussions on hybridity have centered on issues of race and ethnicity. But if East Berlin is conceived as a Soviet-German hybrid and West Berlin as an American-Germany hybrid, Castillo's study demonstrates how the study of hybridity may extend to ideology as well.

In the final chapter, "Porous Boundaries: Fence Patterns and Mexican-American Identity in San Antonio, Texas," Robert Mugerauer describes how the Mexican-American residents of a San Antonio housing project, Mirasol Homes, managed to shape their environment partly through the choices and uses of fence forms, and so articulate a unique Mexican-American identity. Noting that the fences and social uses of complex space both hold to Mexican traditions and embrace American forms and materials, Mugerauer suggests that institutions or authorities which build and manage such environments must be aware of the political possibilities of hybrid forms as they help resolve issues of daily life.

Based on these contributions, the book sets forth the central question of whether hybridity in urbanism can be seen as congruent to its abstract mother concept of "third space" or "in-between" position.

HYBRIDITY AND GLOBALIZATION

I have argued elsewhere that the notion of a "world culture" is problematic. Even if there is to be a world culture, it is a culture marked by the careful and manipulated management of difference and diversity, rather than the replication of similarity or uniformity. Like others, I have also contended that a world culture is essentially a culture of dominant groups, in which the persistent diversity of constituent local cultures is often a product of globalization itself. And I have written that any apparent convergence between the urbanisms of the haves and have-nots may only reflect the self-representation of the dominant particular, and not a true integration of the "third space" into a world system.[41]

Despite the current preoccupation with globalization, the history of the world demonstrates a movement toward cultural differentiation, and not homogenization, in which each individual claims allegiance to more than one culture and invokes different identities at different times and in different places.[42] If in such an interconnected cultural realm, hybridity is accepted as an inherent constituent of national identity, then any particular form of urbanism must be accepted only as a reflection of a specific transitional stage in the life of a society. Indeed, globalization has made the very issues of identity and representation in urbanism cumbersome. Because culture has thus become increasingly placeless, urbanism will continue to be an arena where one can observe the specificity of local cultures and their attempts to mediate global domination. And as the nations of the globalized world order become more conscious of their religious, ethnic and racial roots, I have argued that they will continue to seek forms and norms that represent these subidentities, even if these send confused messages to a global audience, who will ultimately experience them through the space of flows.[43]

This is what I believed a few years ago, and in my mind much of it still applies in the last days of the millennium. As I look at the case studies of the first section of this book, I am convinced that the built environment of the late nineteenth century provides an equally significant arena to witness the emergence of the cultural conflicts of early modernity. The urbanism of that time was, as it is now, an appropriate medium to examine such concepts as hybridity. However, the chapters in the second section demonstrate that globalization has made it increasingly difficult to use urban form to understand the spatial manifestations of human behavior. Likewise they cast doubt on the utility of "urbanism" as an analytical category that may represent the cultures of peoples and places. The belief that space shapes and is shaped by social processes is neither a new phenomenon nor a practice in need of justification. However, the second section taken as a whole attempts to untangle the knot that ties people and place. The connection between a particular culture and the stable terrain it occupies in the form of nations is being challenged through the lenses of ethnicity, heritage and globalization.

Another equally important challenge is the emergence of borderlands as the typical sites of global encounter. These borderlands no longer simply occupy the margin, and they can no longer be seen as stationary topographical sites between two fixed locales (nations, societies or cultures). They are now the "interstitial zone of displacement and deterritorialization that shapes the identity of the hybridized subject."[44]

The example of Hong Kong is a good one here. Quentin Lee has argued that intellectuals who grew up in Hong Kong, and who either remain in Hong Kong or have emigrated to the West, usually identify themselves as Chinese, or as a hybrid, or simply as "someone from Hong Kong." In this scenario, Hong Kong is "a purely geographical location devoid of national boundary and cultural identity," a place in between China and the West "without any national imagining of a common territory, tradition, or origins."[45] It is possible to imagine Mombasa in the seventeenth century or Izmir in the nineteenth century as similar to Hong Kong of today in that their residents may also have seen themselves as "uncompromised by the myth of origins or the necessity to 'perform' or supplement the nation."[46] I accept "nation" here to mean the larger empire, and "origin" to mean the indigenous race. We must remember here that Hong Kong has been a "blind spot" for postcolonial discourse, where the "romanticization of the colonialism-wrecked East" has been deconstructed by its very success.[47] But it is precisely in such unconventional sites that the challenges faced by a "hybrid" society and the complexities posed by a hybrid urbanism appear most clearly.

Katharyne Mitchell's study of Hong Kong immigrants in Vancouver showed how Hong Kong Chinese capitalists who have moved overseas have managed to project their images interchangeably as those of the transnational, the transcultural cosmopolitan, and the "ethnic Chinese." In doing so, they have positioned themselves both on the margins of the Chinese and Canadian nations and "at the lucrative center of Pacific Rim business."[48] It is a flexible position at once on and between the border, as well as within the center.

These examples echo the important work of Gloria Anzaldua who started the borderlands debate with her book *Borderlands/La Frontera*. Anzaldua writes, "The new *mestiza* copes by developing a tolerance for contradictions, a tolerance for ambiguity. . . . She learns to juggle cultures. . . . Not only does she sustain contradictions, she turns the ambivalence into something else."[49] Arif Dirlik has continued this line of thinking and expressed sympathy with the postcolonialists when they evoke "borderlands" as opposed to fixed identities. For him, "borderlands point not to liberated zones but to zones that pose new problems for the task of liberation and may at best serve as points of departure, not as points of arrival." To address this issue, he proposed a "critical localism" to resist the inequalities of both global power and local oppression inherited from the past. "The local has emerged as the site of resistance as capitalism, in its manipulation of local differences, has brought the local to the surface of consciousness."[50] Dirlik has not been unaware of the problems emerging from his affirmation of the local, such as the invocation of local "purity" as an excuse for a reactionary revival of older forms of oppression. But he has rightfully dismissed such localism as "romantic nostalgias" and "hegemonic nationalist yearning of a new kind."[51] Here "the local has meaning only inasmuch as it is a product of the conjuncture of structures located in the same temporality but with different spatialities."[52]

To address the idea of "unity in diversity," Dirlik has drawn on Henry Giroux's identification of what he has called "formative narratives." Because these are capable of analyzing "difference within, rather than against unity," they can "provide the basis for historically and relationally placing different groups or local narratives within some common project."[53] Through the notion of "creolization thought," Ulf Hannerz has examined this relation between the local and the global from another angle. He has observed that it is now more difficult than ever to see the world as "a cultural mosaic of separate pieces with hard, well-defined edges."[54] Instead, he has argued that the flow of culture between countries and continents may result in another diversity of culture, based "more on interconnections than on autonomy."[55]

> Creolization thought is open-ended; the tendencies toward maturation and saturation are understood as quite possibly going on side by side, or interweaving. There may come a time of decreolization, the metropolitan Standard eventually becoming dominant, but a creole culture could also stabilize, or the interplay of center and periphery could go on and on, never settling into a fixed form precisely because of the openness of the global whole.[56]

Creolization itself may be an odd hybrid term. For example, Jan Pieterse has noted that in the Caribbean and North America it stands for the mixture of African and European cultural practices, as in cuisine. Yet in Hispanic America, *criollo* originally denoted people of European descent born in the Americas. The appeal of the word "creolization" partly comes from its tendency to cut against the grain of nineteenth century racism and the accompanying abhorrence of *métissage* as miscegenation. The idea of creolization has thus highlighted "what has been hidden and valorizes boundary crossing."[57]

In contrast, the Latin American term *mestizaje* refers to boundary-crossing mixture itself. Since the early part of the century, this concept has been part of a hegemonic elite ideology, which, in effect, has stressed whitening, or Europeanization, as the overall project for Latin American countries. Latin America would thus achieve modernity, "[as] the European element is supposed to maintain the upper hand, through the gradual 'whitening' of the population and culture."[58] Both creolization and *mestizaje* must be seen in the context of the early development of the Americas. However, it must also be remembered that hybridity is a condition that cannot be viewed independently of the nostalgic tendency which assumes this precolonial purity.

Indeed, hybridity is not new and should not be equated with globalization. As Pieterse has suggested, if one accepts that "cultures have been hybrid *all along*, hybridization is in effect a tautology."[59] However, even though hybridity has always been there, what is new is the recognition on the part of governments and cultural elites that hybridization is a "political and normative discourse."[60]

In his discussion of the Caribbean, Stuart Hall, one of the most articulate scholars of the identity and diaspora discourse, offers an interesting rendering of three "presences": *présence Africaine, présence Européenne*, and the third, *présence Américaine*.[61] According to Hall, *présence Africaine* is the site of the repressed, "apparently silenced beyond memory by the power of the experience of slavery." *Présence Européenne* is about exclusion, imposition and expropriation, which interrupts the innocence of the whole discourse of "difference" in the Caribbean by introducing the question of power. *Présence Américaine*, is not so much about power, as ground, place and territory. "None of the people who now occupy the islands — black, brown, white, African, European, Spanish, French, East Indian, Chinese, Portuguese, Jew, Dutch — originally 'belonged' there."[62] *Présence Américaine* is then the third term and the primal scene of encounter. It is the space where creolization, assimilation and syncretism were all negotiated. Hall has argued that the *présence Américaine* of the "New World" — America, *Terra Incognita* — "is itself the beginning of diaspora, of diversity, of hybridity and difference." Seen from within the third space, I would argue that "*Terra Incognita*" at the end of the twentieth century is becoming the entire globe itself.

The assumption that hybrid environments simply accommodate or encourage pluralistic tendencies or multicultural practices should be turned on its head. Hybrid people do not always create hybrid places, and hybrid places do not always accommodate hybrid people. All that can be hoped for at the beginning of the 21st century are environments that harbor the potential for growth and change and peoples who may find the possibility of adapting and adopting otherness as a legitimate form of self-identification.

In the end, let me return to the fictional characters with which I began this essay. In search for her missing child, Esperanza travels from her humble Mexican village to the rowdy brothels of a border town where she meets her man. Their lives are changed forever in this space of the in-between. But neither Esperanza nor Scott are hybrid people, nor do they care to be. She was working as a prostitute in Tijuana, Mexico; he is a respected all-American judge from San Diego. It is precisely because their identities are especially place bound that the place of their encounter and the space created by it become arenas of "unavoidable frontal collision." Their story points to the emergence of borderlands and the sites of the in-between as the prime and typical spaces of global encounter. But these borderlands are no longer fragments anchored between two fixed and well-defined places. And sites of the in-between no longer simply occupy the margins of the periphery, for they have moved firmly to the center of the core. These will be the new spaces and places of hybridity, and the new locales of postmodern global culture in the next century.

NOTES

1. M.A. Escandón, *Esperanza's Box of Saints*, New York, Scribner Paperback Fiction, 1999, pp.158-159.

2. M. Wald, "Clean Cars Get Real: Makers of Hybrid Vehicles Add Fuel to Electric Mix," *New York Times*, May 29, 1999.

3. E. Rothstein, "From Darwinian to Disneyesque," *New York Times*, July 15, 1999.

4. Ibid.

5. J. Greenberg, "Combining Two Identities: Arab in Israel Parliament," *New York Times*, July 20, 1999.

6. "Identity," *The Oxford English Dictionary*, 2nd ed., CD-ROM, New York, Oxford University Press, 1992.

7. K. Woodward, "Introduction," in Woodward (ed.), *Identity and Difference*, London, Sage, 1997, pp.1-48.

8. S. Hall, "Introduction: Who Needs Identity?" in Hall and P. Du Gay (eds.), *Questions of Cultural Identity*, London, Thousand Oaks, CA, and New Delhi, Sage Publications, 1996, p.4.

9. B. Anderson, *Imagined Communities*, London, Verso, 1983.

10. R. Williams, *Keywords*, New York, Oxford University Press, 1976.

11. It is interesting to note that none of the central terms used in this book — identity, hybridity, difference and diversity — can be found in Williams' book. Among the terms he does invoke are culture, ethnicity, tradition, ideology, and image.

12. "Hybrid," *Oxford English Dictionary*, 2nd ed., CD-ROM.

13. P. Werbner, "Introduction: The Dialectics of Cultural Hybridity," in Werbner and T. Modood (eds.), *Debating Cultural Hybridity: Multi-Cultural Identities and the Politics of Anti-Racism*, London and Atlantic Highlands, NJ, Zed Books, 1997.

14. S. Pile, "Masculinism, the Use of Dualistic Epistemologies and Third Space," *Antipode*, vol.26, no.3 (1994), pp.255-277. The quotation is from p.263.

15. N. Papastergiadis, "Tracing Hybridity in Theory," in Werbner and Modood, *Debating Cultural Hybridity*, p.258.

16. Ibid., p.257.

17. R.J.C. Young, *Colonial Desire: Hybridity in Theory, Culture and Theory*, London and New York, Routledge, 1995, p.6.

18. Ibid., p.7.

19. Ibid., pp.17-18.

20. Ibid., p.19.

21. Ibid., p.6.

22. Ibid., p.28.

23. N. Papastergiadis, "Restless Hybrids," *Third Text* 32 (Autumn 1995), pp.9-18.

24. H. Bhabha, *The Location of Culture*, London and New York, Routledge, 1994, pp.113-114.

25. Ibid., p.116.

26. For a careful analysis of Bhabha's works, see M. Fludernik, "The Constitution of Hybridity: Postcolonial Interventions," in Fludernik (ed.), *Hybridity and Postcolonialism: Twentieth-Century Indian Literature*, Tubingen, Stauffenburg Verl, 1998, especially pp.22-54.

27. "Difference," in the *Oxford English Dictionary*, 2nd ed., CD-ROM.

28. "Diversity," in ibid.

29. Bhabha, *The Location of Culture*, p.34.

30. H. Bhabha, "The Third Space: Interview with Homi Bhabha," in J. Rutherford (ed.), *Identity, Community, Culture, Difference*, London, Lawrence and Wishart, 1990, p.208.

31. Ibid., p. 208.

32. Ibid., p. 209.

33. K. Mitchell, "Different Diasporas and the Hype of Hybridity," *Environment and Planning D: Society and Space*, vol.15 (1997).

34. Ibid., p.534.

35. Ibid., p.536.

36. Ibid., p.539.

37. K. Marx and F. Engels, *The Communist Manifesto*, New York and London, Modern Reader Paperbacks, 1964, p.8.

38. J. Friedman, "The Hybridization of Roots and the Abhorrence of the Bush," in M. Featherstone and S. Lash (eds.), *Spaces of Culture: City, Nation, World*, London, Thousand Oaks, CA, and New Delhi, Sage Publications, 1999, pp.237-238.

39. Friedman, "The Hybridization of Roots and the Abhorrence of the Bush," p.235.

40. J.N. Pieterse, "Globalization as Hybridization," in M. Featherstone et al. (eds.), *Global Modernities*, London, Thousand Oaks, CA, and New Delhi, Sage Publications, 1995, p.51.

41. N. AlSayyad, "Culture, Identity and Urbanism in a Changing World," in M. Cohen et al. (eds.), *Preparing for the Urban Future*, Baltimore, Johns Hopkins University Press, 1996, pp.108-122.

42. T. King, *Culture, Globalization, the World System*, London, McMillan, 1991, p.16.

43. N. AlSayyad, "Urbanism and the Dominance Equation," in AlSayyad (ed.), *Forms of Dominance: On the Architecture and Urbanism of the Colonial Enterprise*, Aldershot, Avebury, 1992, pp.1-26.

44. A. Gupta and J. Ferguson, "Beyond 'Culture': Space, Identity and the Politics of Difference," *Cultural Anthropology*, vol.7, no.1 (February 1992), p.18.

45. Q. Lee, "Delineating Asian (Hong Kong) Intellectuals: Speculation on Intellectual Problematics and Post/coloniality," *Third Text* 26 (1994), p.12.

46. Mitchell, "Different Diasporas and the Hype of Hybridity," p.540.

47. Lee, "Delineating Asian (Hong Kong) Intellectuals," p.12.

48. Mitchell, "Different Diasporas and the Hype of Hybridity," p.551.

49. G. Anzaldua, *Borderlands/La Fronters: The New Mestiza*, San Francisco, Spinsters/Aunt Lute, 1987, p.79.

50. A. Dirlik, *After the Revolution*, Hanover and London, Wesleyan University Press, 1994, pp.107-108.

51. Ibid., p.108.

52. Ibid., p.110.

53. H. Giroux, *Border Crossings: Cultural Workers and the Politics of Education*, New York and London, Routledge, 1992, p.54.

54. U. Hannerz, "The Global Ecumene," in Hannerz, *Cultural Complexity: Studies in the Social Organization of Meaning*, New York, Columbia University Press, 1992, p.218.

55. Ibid., p.266.

56. Ibid.

57. Pieterse, "Globalization as Hybridization," p.51.

58. Ibid., p.54.

59. Ibid., p.64.

60. Friedman, "The Hybridization of Roots and the Abhorrence of the Bush," p.241.

61. S. Hall, "Cultural Identity and Diaspora," in Rutherford (ed.), *Identity: Community, Culture, Difference*, pp.222-237.

62. Ibid., p.234.

PART I

IDENTITY AND TRADITION IN PREMODERN URBANISM

1

Cross-Cultural Currents: Swahili Urbanism in the Late Middle Ages

Thomas R. Gensheimer

Rarely are cities monolithic in form or design, but rather represent the culmination of specific historical events and general cultural influences. Most cities mature gradually, as new buildings are added while others are removed or modified, creating an accumulation of structures which reveal a complicated legacy of growth and change. These changes are at times the result of shifting forms of political status, but more often they are the outcome of influences which derive from the exchange of ideas and efforts to create new forms of identity.

Walking through the East African city of Mombasa today, one can see how waves of events and influences remain frozen in its urban landscape. Approaching downtown along Kilindini Avenue, two monumental elephant tusks flank the street, reminders of the ivory trade on which the city once prospered. The street is broad and lined with multistory structures largely built during the 1930s. In their time these buildings represented something new, reflecting various modern architectural design styles introduced from Europe. Such stylistic importation, however, actually began a decade earlier with the Neoclassical columned and arcaded facades of the buildings farther along, where the name of the street changes to Nkrumah Road. Here, the buildings of institutions such as the National Bank of Kenya and Barclay's Bank make a different statement, and their presence serves as a reminder of the economic dimension of the city's former colonial domination. The *chatri* domes above the entrance to the Oriental Building also make reference to this former colonial status and indicate the extent to which it brought Indian involvement in the economic development of the city. Finally, near the waterfront stand the surviving symbols of more direct forms of colonial political control and domination. The towered facade of the Law Courts is a visible reminder of the English colonial administration, which seized control of Kenya at the end of the nineteenth century. And the massive walls of the early seventeenth-century fortress speak of older colonial legacies: first, by the Portuguese, who sought to dominate the East African coast to extract its wealth and control the trade route to India; and later by Oman, whose forces used the fort as an administrative center after capturing the city from the Portuguese in 1698.

Somewhere across the street, I was told, a remnant of the old town wall survives, which in the mid-nineteenth century, according to the infamous explorer Richard Burton, rose to a height of 10-12 feet.[1] Little remains of this earlier city, except for the small Mandhry Mosque which has stood on its site since 1570/71 (although its conical minaret was probably a later addition). As one wanders through the narrow streets of "Old Town" Mombasa, the most visible reminder of the past is provided by multistoried balconied houses, the design of which was heavily influenced by western Indian traditions. The houses speak of the merchants from Gujarat who were encouraged to settle here during the latter part of the nineteenth century. Many of the massive, carved wooden doors which decorate shop entrances and the entryways of the houses of wealthier merchants were once even produced overseas and imported to East Africa.[2]

In the old city, it is still possible to glimpse evidence of these earlier times. For example, traditional sailing vessels still dock on occasion at the Old Town port. On one occasion I watched a team of workers manually off-load a supply of large wooden beams, and then systematically restock the ship's hull with plastic containers and kitchenware. Two well-dressed businessmen, also interested in this scene, spoke to each other in Arabic until I interrupted them. One was from Yemen but said he had lived all his life in Mombasa. He informed me the ship was bound for Somalia, and that such an event was not unusual, just as it was not unusual to hear Arabic spoken in Mombasa.

During the late Middle Ages, Mombasa was a cosmopolitan trading center, attracting ships and merchants from as far away as southern Arabia and India. Many such foreign merchants made their homes on East African shores. Others would come, stay for a while, and then go with the monsoon winds, leaving their goods and taking their profits. As unintended cargo, these foreign merchants brought new ideas and influences to the East Africa coast, which helped shape the developing urban environment in medieval times. The process was much the same as that by which Western and Indian influences molded Mombasa during its expansion in the nineteenth and early twentieth centuries. But unlike the later case, the medieval foreign influence did not derive from colonial domination, but rather from a two-sided economic and cultural interaction, which molded urban development and resulted in the blending and adaptation of architectural styles. The process was similar to the introduction of the modern buildings along Kilindini Avenue — which brought new concepts of architectural design to the city. These new buildings, according to Maurizio Sabini, "never marked . . . a total break with the past," because their mix of commercial shops on the ground floor and residences above followed the pattern of traditional houses of earlier times.[3] Likewise, the elaborately carved wooden doors which frame the entranceways of historic houses and shops in Old Town Mombasa, derive from an architectural tra-

Figure 1.1. Bajuni-style carved wooden door, from Lamu, Kenya. (Photo by author.)

dition common throughout the cities of the western Indian Ocean in earlier times. On the East African coast, these doors remained a vibrant tradition, developing their own decorative styles which reflect aesthetic preferences linked to regional and ethnic differences.[4] And despite their possible foreign origins, such doors have become a characteristic feature of traditional Swahili architecture, serving as an identifying feature of Swahili culture to this day **(Figure 1.1)**.

IMMIGRATION AND THE ORIGINS OF SWAHILI CULTURE

When the Portuguese arrived in East Africa at the close of the fifteenth century, Mombasa was already a major city, described as containing "edifices . . . of stone and mortar, with windows and terraces like those of Spain."[5] It was just one of many urban centers the Portuguese encountered along the coast of East Africa, from Mozambique to southern Somalia. The Portuguese regarded these cities as wealthy urban centers ripe for plunder and colonial domination. Thus, of Dom Francisco d'Almeida's voyage to Kilwa in 1505, the Portuguese historian De Barros wrote, "From our ships the fine houses, terraces, and minarets, with the palms and trees in the orchards made the city look so beautiful that our men were eager to land and overcome the pride of this barbarian."[6] Mombasa, too, was attacked and looted during d'Almeida's campaign, and Dom Francisco ordered his soldiers to set fire to the city, "destroy(ing) the greater part of this city of abomination."[7]

To the Portuguese, the Swahili of the East African coast were little different than their familiar enemies in other parts of the Islamic world. They described as Moors "all [who] speak Arabic," and "dress very well in the Moorish and Turkish fashion."[8] For the Portuguese, city residents' knowledge of Arabic, the common religion of Islam, and the similarities in architecture and material culture with Arab peoples could only be interpreted one way. They concluded that these African cities were colonial outposts of Islamic societies from the Persian Gulf or Southern Arabia — a notion supported by local oral histories, which often traced their residents' ancestry to early waves of Arab or Persian immigration. Yet such notions relied on the belief that East Africa had been a cultural vacuum, awaiting occupation by superior cultural groups from the more civilized lands to the north and east. Such ideas reflected a colonial attitude toward African cultures, and would become a red herring that would wrongly influence scholarship on Swahili culture and cities for many years.[9]

Linguistic and archeological studies have now put to rest the notion of "Arab" or "Persian" colonization of the East African coast.[10] The studies indicate the independent development of a coastal culture which over time distinguished itself from other African cultures through the exploitation of coastal resources and the adoption of a maritime way of life. Yet a significant feature of such a coastal life had always been the regular presence of foreign traders who came to East Africa since Indian Ocean trade was revived under the Abbasids. Traders from the Persian Gulf came to Africa to find slaves, who were in high demand in the agricultural fields of southern Iraq and Bahrain, and to find a source of hardwood, particularly mangrove poles, which were needed for the extensive building campaigns in the largely treeless regions of the Gulf.[11] Then, with the rise of Fatimid power at the end of the tenth century, trading patterns shifted to emphasize Red Sea destinations, and the demand for African goods changed, with luxury items such as rock crystal and ivory becoming highly prized in the workshops of Fatimid Cairo.[12]

A significant feature of this trade was the fact that sailing patterns on the Indian Ocean required significant layovers in distant lands, as foreign merchants awaited shifts in monsoon winds to make return voyages possible. And, eventually, some merchants stayed as a means to facilitate this trade, settling within existing communities and dealing with local merchants who could take advantage of existing regional trade networks. Such instances would most likely have been isolated events at first, and, given the modest size of early Swahili communities, such settlers would have been readily assimilated. Yet, despite its modest beginnings, such a pattern of foreign impact on the development of Swahili culture and settlement patterns eventually gained significance, first through the introduction of Islam, which became the dominant religion of the coast by the eleventh century, and then through a continuing pattern of cultural influences and the introduction of new architectural ideas.

CROSS-CULTURAL INTERACTION DURING THE LATE MIDDLE AGES

The Swahili cities that so impressed the Portuguese were largely a development of the late Middle Ages, when economic forces and historical events contributed to the prosperity and development of the area. At the time, interregional trade formed the basis of urban development throughout the western Indian Ocean, where coastal cities functioned as emporia facilitating maritime commerce. The creation of emporia cities owed much to the cyclical monsoon winds, but such cities also arose out of the needs of wholesale traders to develop central locations for markets and services to maximize profits and deal effectively with the inherent uncertainties of seafaring.[13] As a result, cities such as Aden, Suhar, Siraf, Hormuz, Cambay and Malacca developed as nodes in overlapping networks of maritime trade reaching from Europe to China.

In the late Middle Ages, trade throughout the Indian Ocean expanded greatly, tying the commercial affairs of the Far East and South Asia ever more closely to the Islamic and Mediterranean worlds. One result was that the East African coast, formerly peripheral to the main trading network, became increasingly connected to the commercial and cultural affairs of the western Indian Ocean. As commercial activity in the Indian Ocean grew, by the thirteenth century trade had shifted from primarily luxury goods to more diversified forms, which included bulk commodities and manufactured goods.[14] East Africa profited from this change, because in addition to rare and exotic goods, many of its exports were unfinished raw materials. Gold also became an export of particular interest. Arab traders had been aware that gold was available on the East African coast since the tenth century, and possibly even earlier.[15] Its sources were the mines of the Zimbabwe plateau, and for centuries the settlements of the coast had served as the conduit by which this gold reached world markets. Trade in gold most likely became a significant component of the East African commerce in the latter half of the thirteenth century when worldwide demand increased, Italian city states began to issue gold coins, and economies from Delhi to Egypt became increasingly monetarized.

Such changes in the scope and scale of the Indian Ocean trade during the late Middle Ages dramatically increased the degree of commercial and cross-cultural contacts between East Africa and the larger Islamic world. During the thirteenth and fourteenth centuries, the dominance of Indian Ocean trade by the powerful Karimi merchants, centered at Aden, renewed the power of cultural influences emanating from southern Arabia and spreading around the Indian Ocean basin. The thirteenth century also brought the beginning of a period of migration by Hadrami and Yemeni Arabs, some of whom were known to have come to East

Africa.[16] Among these immigrants were many learned holy men and descendants of the prophet. These men were readily accepted by Swahili society, frequently attained honored positions in the governments of the communities where they settled, and had descendants who assumed leadership roles in existing political institutions.[17] Immigrants may have also come to East Africa from other parts of southern Arabia and Asia, settling in coastal cities and becoming assimilated into the social structure of Swahili settlements. Traditional histories often refer to such patterns of immigration by foreigners, including the Shirazi from Persia at Kilwa and the the Nabahani from Oman at the city of Pate. The presence of Indian merchants in the Swahili cities of the fifteenth century was also noted in accounts by the Portuguese. Their presence reflected a significant realignment in trade, as the Karimi merchants lost their monopoly control of the spice trade in 1429, and Indian Ocean trade became dominated by Gujarati merchants, with Cambay as the principal trading port.[18] The extent of Indian involvement in East Africa can also be seen in the influx of Africans to western India during the mid-fifteenth century. Although originally brought as slaves, many rose to prominent positions in the military or administrative organizations of the sultanates of the Deccan, Bengal and Gujarat. The dynasty of the sultans of Janpur may similarly have descended from African slaves brought to India in the late fourteenth century.[19]

It should also be noted that Swahili merchants were not simply passive participants in the Indian Ocean trade; many were actively involved in travel and seafaring. An eyewitness account by Tome Pires in the early sixteenth century stated that many foreign merchants accompanied the Gujarati in their ships during their voyages to Malacca, specifically noting merchants from the East African cities of Kilwa, Malindi, Mogadishu and Mombasa.[20] Similarly, it is well documented that ambassadors from the Sultan of Malindi were received in the court of Yung-lo, the third Ming emperor of China, in 1415, precipitating the large seafaring expedition of the eunuch admiral Cheng Ho in 1417-1419 which reached Malindi on the central Kenyan Coast. Such extensive commercial and cultural contacts would have had a significant effect on the development of Swahili civilization in the late Middle Ages, contributing to urban development and providing influences from which a distinctive Swahili architecture and urban form would emerge.

KILWA IN THE FOURTEENTH CENTURY

Relatively little remains of the extensive late-medieval urban settlement along the coast of East Africa. However, there is evidence in the fourteenth century of urban expansion in the area during a period of increasing reliance on imported goods — often to the detriment of local industries.[21] During this time, the city of Kilwa, located on an island off the central coast of Tanzania, became the most prosperous urban center in East Africa. Ibn Battuta, who traveled there in 1331, described Kilwa as the "principal town of the coast." Its rise to prominence may have been due to its residents' ability to gain control of the gold trade, as recorded in its traditional history. The city was described as built largely of reeds, which may refer to the large number of structures built of impermanent materials, probably timber and thatch. Yet two of the grandest architectural achievements of the East African coast were also found there during the time in Kilwa: the Great Mosque and the palace complex of Husuni Kubwa.

The Kilwa Great Mosque began as a small, flat-roofed structure built of roughly squared coral blocks laid in courses, a technology that originated on the shores of the Red Sea in early Islamic times.[22] Around 1300 A.D. a southern extension was added, enlarging the prayer hall five-fold, creating the largest mosque in medieval East Africa, and reflecting the rapid expansion of the city. The addition was built using coral rag (rubble stone) set in lime mortar without distinct courses, a construction method that became widespread in subsequent centuries. A novel method was also used to cover the addition, consisting of a series of multiple domes or vaults supported on a series of monolithic columns carved from fossilized coral.

As fine as this mosque may have been as an architectural statement, it was in the construction of the palace complex of Husuni Kubwa that Swahili builders most excelled in such new construction methods, creating the largest and most monumental structure in all of medieval East Africa. Located outside the city itself, the complex consisted of two distinct structures, separated by a gully. One of the structures, Husuni Ndogo (literally translated as the little fort), consisted of a large rectangular enclosure with an undifferentiated internal space. The design was unique and enigmatic, but had an implied defensive character in the thickness of its walls, its single protected entrance, and the presence of multisided bastions which buttressed its corners and sides. The other structure, across the gully, was Husuni Kubwa, a palace of great scale and grandeur. It took the shape of a complex arrangement of formal courtyards and domestic quarters, built to provide the sultan with appropriate spaces for public audiences; living spaces for his wives, family and servants; and facilities to conduct commercial activities on a grand scale. The sultan's private quarters were at the northern end of the palace, with rooms organized around a rectangular sunken court. Adjacent was a chamber with a large octagonal pool, an architectural feature without precedent in East Africa. To the south was an impressive audience court, built as a large sunken space with nine monumental steps leading to a sheltered pavilion. It was here that the sultan would receive guests and conduct public affairs. To its rear was a enormous open courtyard surrounded by storerooms to which access was limited, which provided a center for the sultan's commercial activities.

THE FLORESCENCE OF SWAHILI URBANISM

Despite these impressive architectural achievements at Kilwa, it was during the later part of the fourteenth and throughout the fifteenth century that Swahili cities developed their distinctive urban character. During this time, as masonry construction became widespread, new architectural forms were developed to accommodate social changes and the new urban institutions that resulted from increased trade and cultural contact. All along the coast, new cities were founded, and established settlements were expanded significantly with the addition of new public structures, the extensive remodeling of older ones, and the increase in number, size and complexity of domestic structures.

Swahili cities of the fifteenth century reflected their position as centers of interregional seafaring trade. The location of the Swahili coast at the meeting point of the western Indian Ocean trading network and the inland trading routes of continental Africa led to the siting of cities where they could readily accommodate the transshipment of goods. Swahili cities of the period thus were consistently located with some consideration for access from the sea, either on small offshore islands or along the coastal inlets which are a common geographical feature of the area. Often such inlets had only shallow channels that allowed access to the settlement only at high tide, and provided a protected harbor where boats could be sheltered. At such sites, specific harboring facilities were never constructed, because traditional docking involved careening on shallow beach fronts for loading and unloading during low tide. Sea walls were occasionally built, but these served largely to reclaim land, provide building platforms, or prevent erosion by the sea.

Unlike at Kilwa, where Husuni Ndogo may have served the defensive needs of the community, by the fifteenth century most Swahili cities were surrounded by walls. Such walls were relatively short (generally less than 6 feet high), and frequently followed an irregular outline. Square rooms were often built along the walls, forming gate-like structures which added to their defensive character. At Gedi, one of the largest ruined cities on the central coast of Kenya, the city wall was more than a mile in circumference, yet only three gates were found, indicating a great restriction of access to the city.

The relatively low height and irregular placement of city walls suggested they were also constructed to demarcate social boundaries. In fact, evidence of often-extensive communities outside the walls indicates certain segments of the medieval Swahili urban community were specifically excluded by them. With the exception of mosques and tombs, these outlying communities were built exclusively of impermanent materials such as mud, timber and thatch, in contrast to the many domestic structures built of stone within the walls. In later Swahili settlements, stone construction served as a symbolic indicator of social status, and often demarcated a division of the community into two *mikao*, or moieties. Such a division may also have existed in medieval Swahili settlements, indicating an urban community divided between high-status elites (in later times known as the *waungwana*) and lower-status commoners (the *wazalia* or *washenzi*). In fact, the presence of such elite groups today may have originated in medieval times, and may have been related to the emergence of stone houses.[23] Elites generally rose to prominence in the late Middle Ages through their involvement in the export sector of the Swahili economy, serving as middlemen, and forming an oligarchy of merchants "which monopolized most of the surplus" and led to "a life of ostentation and luxury."[24] In medieval Swahili society, therefore, the city wall, along with differing construction materials, may have served as a means of maintaining social boundaries, creating an architectural manifestation of existing social divisions.

Like many Islamic cities around the Indian Ocean, the Friday, or *Jami'a*, mosque was the central institution of the Swahili city, and generally its most monumental architectural undertaking. Even in the smallest settlements, the importance of this mosque was accentuated by its superior construction, often being the only structure built of stone. In larger cities, the Friday mosque was often built on a grand scale to accommodate the entire Muslim community. Yet despite their importance, such mosques did not exhibit monumental facades or towering minarets designed to announce their presence in the urban landscape. In fact, except for a few early examples in southern Somalia, minarets were never built on medieval Swahili mosques. Instead, the call to prayer was conducted from the roof, which was reached by a simple staircase from an outer court.

In plan, the medieval Swahili Friday mosque was unique in many ways **(Figure 1.2)**. In particular, it lacked a courtyard or central open space, which

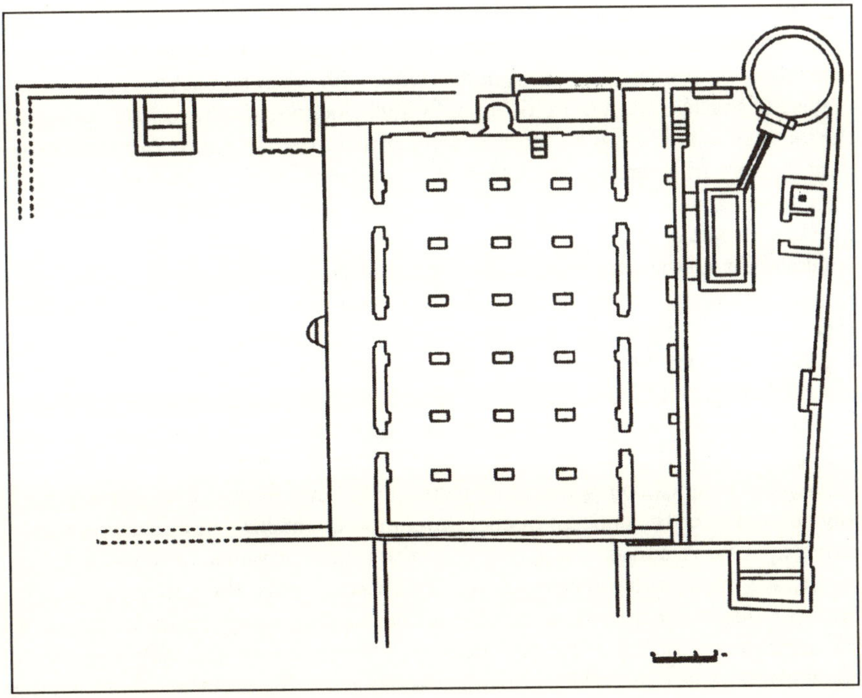

Figure 1.2. Plan for the Friday Mosque at Gedi in the late fifteenth century. (Based on James Kirkman, *The Arab City of Gedi: Excavations at the Great Mosque*, London, Oxford University Press, 1954, fig.1; and Peter S. Garlake, *The Early Islamic Architecture of the East African Coast*, Memoir no.1 of the British Institute of History and Archaeology in East Africa, London, Oxford University Press, 1966, fig.24.)

defined the typical Friday mosque in most Islamic cities of the time throughout the western Indian Ocean region. Rather, the entire space was enclosed, creating an elongated, rectangular prayer hall divided into narrow bays by rows of broad, closely spaced columns. In many cases, a central row of columns even blocked the *mihrab* aisle, obscuring the decorated niche in the *qibla* wall. The main space was often further divided by the addition of verandahs to either side that may have functioned as transitional spaces leading to the main *musalla*, or prayer hall, but also provided additional spaces for prayer. Decoration within the mosque was largely restricted to the *mihrab* niche, which was often accentuated by multiple rebated arches placed within a rectangular architrave. Sharp, detailed edges were also created by individually carved and fitted coral blocks which would be covered in white lime plaster. In addition, inset porcelain or glazed bowls were often added within the spandrels, flanking the spandrels, or within the *mihrab* niche itself. Such a decorative system was unique to East Africa, accentuating the *mihrab* by the addition of color and luster to its plastered surfaces.

In addition to the Swahili city's main mosque, smaller mosques formed additional ritual centers, and were located so as to provide neighborhoods with places for daily prayer and public affairs. Small community mosques were a common feature of historic Swahili cities, and they were often constructed to serve the needs of specific ethnic groups or other communities within the urban population. The proliferation of smaller mosques in medieval Swahili cities was mostly a characteristic of the fourteenth and fifteenth centuries, and may have been indicative of a corresponding increase in heterogeneity within the urban population, a result of expanding commercial and cultural contacts. These communities may have formed individual quarters, or *mitaa*, as they are known in surviving Swahili towns, whose residents generally shared the same social status, clan affiliation, or ethnic origin.[25]

A further important element of the medieval Swahili city were its monumental tombs. Commonly, such tombs were located near mosques, often outside the *qibla* wall, but burial enclosures and monumental tombs were also regularly placed throughout the city, often standing as individual monuments or grouped into small clusters in open areas of the city. In other cases, tombs could be associated with domestic structures, as at Gedi where they formed the main feature of the palace annex courtyard. The form of medieval Swahili tombs was generally derived from the concept of enclosure, monumentalized in different ways — either by massive construction, covering with domes or vaults, or the addition of decorative features or superstructures. Most commonly, this involved the creation of decorative panels along the eastern facade, using precisely cut coral blocks to create multiple recessed surfaces or decorative arches. In many cases, a superstructure was also added which stood out above the enclosure walls, adding to the tomb's visual impact and making it recognizable in the urban setting. Such a strategy often also involved the addition of a rounded slab or tall pillar to the eastern facade. These pillars could be rounded, multisided or fluted, and they were often decorated with imported porcelain or glazed bowls. They formed a distinctive building type unique to the medieval East African coast **(Figure 1.3)**.

Figure 1.3. Monumental pillar tomb in the courtyard of the Gedi Palace annex, fifteenth century. (Photo by author.)

The elaborate construction of such monumental tombs suggests that they belonged to rulers or influential religious leaders, and that they were built as reminders of the piety or status of the deceased. They also likely served as symbols of a common history and ancestry, connecting the past and present with visible representations of the continuity of the urban community. In addition, many such tombs may have served as centers of ritual. Even today certain tombs are known as the burial sites of *sharifs* or other holy men, and attract visitors and pilgrims who pray or place special offerings there in the belief that the deceased can provide spiritual assistance. In medieval times the use of monumental tombs as focal points for prayer and offering was reflected in their design, as many had entrances to the burial chamber within. In other cases, such an entrance may have been reduced to a symbol, represented as an arched opening on the eastern facade, to facilitate the placing of offerings or the burning of incense at the tomb, or to focus attention during prayer.

Another significant feature of larger medieval Swahili cities was the palace structure. As seen at Kilwa, the scale of construction reflected the extent to which political and economic control had become centralized by the early fourteenth century. Yet in later times, palaces were largely reintegrated into the matrix of domestic stone structures, representing the centripetal forces in Swahili politics and the social changes brought about by increased wealth and unequal trade. The impressive size and scale of Husuni Kubwa was in many ways, therefore, not characteristic of Swahili cities of the fifteenth century. In particular, its location outside the city limits was atypical for later palace structures. And such architectural features as its octagonal bathing pool and large commercial courtyard were never again attempted by Swahili builders. Yet Husuni Kubwa and other palace structures did share a sequence of formal spaces which appear to have been an integral part of the rituals of Swahili courts. These were described by Ibn Battuta following his meeting with the Sultan of Mogadishu. Typically, a formal monumental entrance provided a setting for people to gather prior to entering. This entrance led to an outer court, which served both as a waiting area for visitors prior to greeting the sultan and a place where they would return to share a meal outside the sultan's presence. A large sunken court usually adjoined this room, and was used for royal audiences and ceremonies. At such events, guests and officials would take designated positions around the sultan, who sat in a special room or on a dais at one end. The inner rooms served as the "private apartments" to which the sultan retreated, leaving his private secretary, *qadi*, *wazirs*, and chief *amirs* to hear complaints and make rulings, consulting the sultan only when necessary **(Figure 1.4)**.[26]

Palaces were not the only representations of political authority in the medieval Swahili city. The growing wealth and economic power of middlemen and merchants resulted in the creation of a new urban elite who competed with the court for political control. Therefore, in addition to their references to kings, sultans and sheikhs, the early historical accounts suggest a diffused power structure in some medieval coastal cities. This type of political system was more republican in nature, and may have been the legacy of an earlier form of political organization where decisions were made by a council of elders.

Medieval Swahili urban form may partly have been derived from the way these two political forces were reconciled within the urban setting. At Kilwa in the early fourteenth century, the palace formed the dominant feature of the urban landscape, reflecting the extent to which the sultan controlled the political and economic activities of the city. But Husuni Kubwa was exceptional in that it was conceived and constructed during a single building campaign — unlike later palaces which often evolved from simpler forms following standard stone house designs. Likewise, the unique nature of the vast commercial courtyard at Husuni Kubwa indicates the Sultan of Kilwa may have been directly involved in controlling interregional trade. This pattern changed in later medieval times, when trade came to be controlled by a class of merchant elites and the sultan came to derive his wealth simply from the collection of customs duties, as the Portuguese

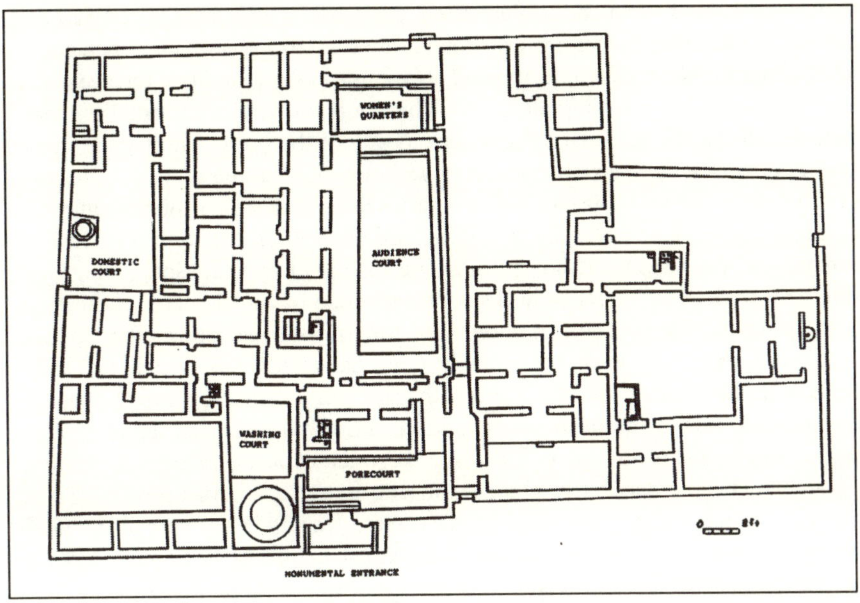

Figure 1.4. Plan of the Palace of Gedi in the late fifteenth century. (Based on James Kirkman, *Gedi: The Palace*, The Hague, Mouton and Co., 1963, p.37, fig.2.)

observed in the sixteenth century.[27] Thus, in the sizable cities of the fifteenth century, palaces came to be built more modestly in scale and within the city, where they competed with many other large stone houses. The numerous stone houses of the time reflect the growing political and economic power of the merchant class and the wider distribution of wealth among the prominent clans.

As the previous discussion suggests, the character of the medieval Swahili city was largely determined by its domestic architecture, in the form of its many stone houses. Archeological remains indicate the vast majority of such houses were single story, despite Portuguese accounts which mention the prevalence of multiple-storied houses.[28] These stone houses were flat roofed, and they enclosed private compounds centered on an internal court. In addition, due largely to practical climatic concerns, houses were consistently oriented to the cardinal points, because the placement of the interior courtyard to the north allowed sunlight to penetrate the rooms to the south which consistently lacked exterior windows. This arrangement both allowed for maximum warming during the coolest months of the year and minimized the sun's penetration into the house during the warmest months. The northern orientation also provided maximum protection against the strong southwest monsoon winds, while allowing breezes into the house from the northeast during the hottest months.[29]

Stone houses generally followed a standardized plan and design. At a large city such as Gedi, which likely contained more than one hundred stone houses, their design was formal and sophisticated, providing a high level of comfort and prestige. The entry spaces of most stone houses led to a rectangular open courtyard, generally sunken into the ground, forming a low surrounding terrace which may have served as a seating platform. This court formed an intermediate space between the public realm and the inner recesses of the house where male guests could be received and affairs conducted without entering the more private parts of the house. Behind this court were a series of long narrow rooms (commonly three in number), arranged back-to-back and entered on the long side through ornamental doorways. Although these rooms could be quite long, they were rarely more than eight feet wide, a span determined by the diameter of the mangrove poles used in roof construction. The innermost rooms were often divided, and small chambers were occasionally built between them, which are today believed to have served as storerooms with trap doors located high up on the wall.

Efforts were made in the construction of Swahili stone houses to create a setting of elegance and luxury. Doorways between rooms were often accentuated with niches or decorative arches, and entrances may have been fitted with carved wooden doors. Amenities such as lavatories and baths were also included within the house, which required the construction of large underground chambers. In the House of the Dhow at Gedi this corbelled chamber was 27 feet deep, and carefully revetted and plastered.[30] Similar sumps were often embedded in the floors of the court and main rooms, allowing for drainage and easy cleaning. Some larger houses had individual wells in their courtyards or arched alcoves faced with carved coral blocks for the elegant storage of drinking water. Wall surfaces were coated in a white lime plaster, at times elaborated with delicate decorative niches or covered with imported textiles hung from pegs in the walls, enhancing the comfort and refinement of these homes.

As they clustered together to form large blocks in the city, these stone houses shaped the urban environment. One house block excavated at Gedi contained at least eleven individual dwellings units, which shared common walls and featured complex interlocking plans. A large city would comprise many such blocks, forming clusters of stone houses which were often concentrated in the center of the city. This urban core would contain other stone structures, including small mosques, monumental tombs, and carefully constructed wells. The total effect was of a tight interwoven network of contiguous walls and narrow passages, where streets, even thoroughfares, were rarely wider than six feet. Such a width would have provided shade from the tropical sun and emphasized the pedestrian nature of transportation in a society where goods and materials were moved largely by human labor. The consistent orientation of stone houses was reinforced by the northern orientation of the *qibla*, creating an orthogonal pattern of streets which extended throughout the settlement.

Away from the urban core, the stone buildings of Swahili cities were more spread out, occurring as isolated structures or, more often, as interlinked clusters. Between them were undoubtedly structures constructed of timber, mud and thatch, much like those of the communities outside the city walls. Thus, Hans Mayr, an eyewitness to the destruction of Mombasa in 1505, recalled, "This city has more than 600 wooden houses covered with palm . . . and the houses built of masonry have these others amongst them." These timber structures were so common that when the Portuguese set Mombasa ablaze, "the whole city burned like one huge fire that lasted nearly all night."[31] From these descriptions one can deduce that nonmasonry structures were an integral component of the medieval Swahili city, although relatively little can be determined today about their nature. Some evidence suggests that such houses were rectangular and that their construction was based on a series of small wooden poles set into the ground. To this framework of vertical members were tied horizontal cross-members, creating a lattice that was filled with mud and coral chips. Roofs may have been peaked and constructed of a pole frame, on which a thatch of coconut fronds could be attached.

Such a dual architectural tradition was a distinctive feature of the medieval Swahili city, and defined a tradition that continues in coastal settlements of the region today. Whatever their specific construction and plan, the effect of the dual tradition on the urban environment was significant, forming regions of lower density within the city and areas of less-regularized planning. Within these outlying areas, there would have been open spaces for gardens, livestock, orchards and palmaries, giving the medieval Swahili city a somewhat rural character, as its boundary tapered into the countryside, only loosely defined by its wall.

THE INDIGENOUS ROOTS OF SWAHILI URBANISM

Traditional societies and their architectural forms are often seen as the product of isolated and independent developments, the result of restricted communications and insular conditions. Yet few societies exist in isolation, and external factors can play a significant role in the creation of new traditions, which emerge from the interplay of available influences and the identity a community seeks to maintain.

Swahili urbanism of the late Middle Ages emerged from two continuing trends. The first was the development of communities throughout the interconnected regions of the Indian Ocean, which due to repeated commercial and cultural contacts came to share similar aspects of material culture. Medieval Swahili communities were inextricably linked to the economic and cultural affairs of surrounding regions, as they purposely sought larger forms of affiliation based on a common Muslim identity and ancestral ties linked to Arab and Asian immigration. Yet the diffusion of foreign influences to the East African coast engendered a uniquely African response, and at no time did such influences displace indigenous traditions. Swahili cities were always an African phenomenon that responded primarily to local needs and indigenous economic and sociopolitical developments. Their architectural and urban forms developed within the historical and cultural context of East Africa, where the urban setting functioned to unite the various groups that coexisted in coastal settlements.

Medieval Swahili cities reflected both indigenous and global forces as the intensification of interregional trade fostered local trade connections within the coastal region and beyond to funnel goods from the interior of Africa to the increasing number of entrepôts and feeder ports. Increasing wealth and competition in turn contributed to the need for defense, and the widespread appearance of walls with gates and other protective structures may have resulted from the diffusion of architectural influences from overseas. In particular, this may have been the case at Kilwa in the fourteenth century, where the design of Husuni Ndogo followed a plan and form that was comparable to defensive structures in areas of coastal Oman, known as *sur* (pl. *aswar*). However, the defensive function of walls was somewhat compromised in later Swahili coastal cities, because their design and placement also served as a symbol of social differentiation and a means of reinforcing internal social divisions.

The presence of the mosque as a central institution and dominant feature of the Swahili city may owe much to the influence of foreign traders. In particular, the Swahili mosque, with its fully enclosed prayer hall and staircase minaret, may have been adapted from small community mosque types of the Persian Gulf region.[32] The use of such a building type as a Friday mosque was unique, however, and reflected a long history of indigenous development, which led to a form which may have better accommodated the climatic conditions of a coast where two seasons of monsoon rain made an open courtyard impractical. The proliferation of smaller mosques may have been indicative of growing ethnic and other social divisions within the late medieval Swahili city. Yet the development of *mitaa*, or wards, was not a pattern adapted from abroad, but rather represented a continuation of settlement patterns which could be traced to pre-Islamic times when different clans came together to form a common settlement.[33]

The construction and significance of monumental tombs shared a common tradition with Islamic Asia and Arabia based on the notion that the Prophet, angels and holy men could function as intermediaries between God and ordinary believers. Their presence in Swahili cities may have represented a continuation of this popular tradition, which may have had its roots in a period of increased migration of *sayyids* and *sharifs* from the Hadramawt in the thirteenth century. However, the integration of tombs into the urban context and their association with domestic structures was particular to Swahili cities, and may have represented indigenous attitudes and practices toward the spirits of the ancestors. In certain coastal societies, spirits were often believed to reside in graves, small huts, or other special sites within the settlement and require offerings and sacrifices to be appeased. Important community members were also traditionally buried within the village itself (in some cases near the house where the deceased lived), and sometimes the burial place was marked with vertical slabs of stone or wooden posts.[34] The paneled facades and pillared superstructures of Swahili monumental tombs were unique to East Africa, as was their decorative treatment by means of the insetting of porcelain or glazed bowls. Overall, the design of Swahili monumental tombs was developed to accommodate the specific symbolic, ritualistic and aesthetic functions of medieval coastal society, fulfilling both African and Islamic ceremonial requirements and spiritual beliefs.

Likewise, palaces and the large stone houses represented social and political institutions particular to medieval Swahili society. Their presence denoted the dual focus of politics in the Swahili city: that of the centralized authority of the sultan and that of the decentralized power base of the council formed by the urban elites. In the early fourteenth century, the sultan dominated both the politics and commercial activities of the city-state, and the design of Husuni Kubwa reflected both its ceremonial and commercial functions. This early palace may have owed many of its architectural features to foreign prototypes or influences. In particular, its dwelling units have been compared with an Umayyad or Abbasid *bayt*, and its great commercial court is today believed to "bear some resemblance to a Persian or Mesopotamian serai."[35] Similarly, the octagonal bathing pool may have reflected Indian or other foreign influences. However, Husuni Kubwa and other Swahili palace structures were likewise designed to accommodate ceremonies and rituals particular to the Swahili coast. The arrangement of a formal entrance, outer and inner courts, and rear apartments redefined the palace in local terms as part of the regalia associated with political authority. Likewise, the arrangement of living spaces in the palace followed the standard plan for Swahili stone houses and had a long history of indigenous development.

By the fifteenth century, the design and grandeur of stone houses served both economic and social functions, because formal market structures with specialized building types were virtually unknown in medieval East Africa. Such a condition reflected both the particular nature of interregional trade in the area and the particular economic institutions operating in Swahili cities. Although traditional markets which served the needs of local communities were generally conducted in makeshift stalls, and may even have lacked any structure at all, overseas commercial exchange was inextricably embedded in social processes. This was because the acquisition of foreign goods served largely to maintain status relationships rather than to fulfill the material needs of the community. This trade was conducted as part of a social contract and did not function in accordance with market principles. Arriving at Mogadishu, Ibn Battuta wrote how his ship was greeted by a crowd of young men carrying food, who presented it to the merchants, each calling out, "This man is my guest." The merchant was then taken into the house of his host, who " sells for him what he has brought and makes purchases for him. Buying anything from a merchant below its market price or selling him anything except in his host's presence is disapproved of by the people of Mogadishu."[36]

Because the conditions of interregional trade were largely determined by social relationships, independent market institutions and formal commercial structures were not needed to facilitate exchange. Instead, commercial affairs were carried out in the stone houses, with domestic architecture developing specific features to accommodate these activities. The vast commercial wing of Husuni Kubwa may have been the grandest example of such an accommodation. In the stone houses of the late fourteenth and fifteenth centuries, the inner storerooms served as specialized chambers for the storage of goods, and the large sunken courts provided space for commercial and social interaction. The occurrence of commercial activities in the stone houses eliminated the need for specialized structures, such as the *suq*, *khan* or *qaysariyya*, which were characteristic features of other medieval cities of the Islamic world.

HYBRIDIZATION AND THE SWAHILI CITY

Commerce between East Africa and the larger world was an unequal trade relationship which functioned to underdevelop the economies of Swahili cities, making them extensions of the highly developed commercial centers of Islamic Asia and Arabia. Within this system, Swahili cities functioned largely as producers of raw materials and consumers of manufactured goods, leading to a dependence on trade and the decline in local industries. Goods imported into East Africa were to a large extent luxury items, such as Chinese porcelains, glazed Islamic ceramics, glass vessels and beads, finely woven cloth of silk, and linen and cotton (at times embroidered with gold and silver threads). These imported goods were acquired and exchanged largely among the affluent elite, serving to reinforce their status and perpetuate social hierarchy through the possession of exotic goods and the emulation of the preferences and consumption patterns of the larger Islamic world.[37]

The linking of societies in global economic systems often leads to the flow of cultural influences from the regions of commercial dominance to less developed regions, resulting in a degree of cultural homogenization. The increasingly global economic developments in the Indian Ocean during the late Middle Ages led to the development of a pre-European "world system" which linked economies from Europe to the Far East in overlapping networks of trade.[38] The connection of Swahili cities to this global trading network contributed to a process of cultural exchange involving the flow of cultural influences from the core metropolitan centers of southern Arabia and Islamic Asia to the peripheral regions of the East African coast. The impact of these cultural flows on the development of Swahili cities can be seen in the development of building types common to Islamic cities around the Indian Ocean basin, including mosques, palaces, monumental tombs, city walls with defensive structures, and domestic structures which were flat roofed, internally focused, and organized by an inner court.

Yet such global processes also promote regional developments. Rather than resulting in the replication of the dominant culture, globalization contributes to the creation of a "global melange" whereby regional societies develop their own distinctive cultural forms.[39] Thus, despite certain similarities with the larger Islamic world, Swahili architectural and urban forms represented regional developments which consistently reflected local functions, decorative styles, and spatial preferences. These indigenous developments were evident in a number of characteristic features: the siting of cities to accommodate indigenous methods of harboring; Friday mosque plans with an enclosed rectangular prayer hall; monumental tomb forms which are without precedent in Africa or Islamic Asia; the design and arrangement of ceremonial and living spaces in Swahili palaces; the economic functions of stone houses; and the consistent orientation of structures so as to create an orthogonal regularity to the urban fabric.

Regional cultural forms that develop from globalization are the result of a process of hybridization, which emphasizes the mixing of cultures rather than their separateness.[40] The medieval cities of the Swahili coast arose from such a hybridization process, as foreign architectural concepts were adapted to the cultural context of East Africa as part of social and economic changes in coastal society. Thus, the medieval Swahili city did not emerge from a simple adoption of foreign styles and building types, but rather from a process by which new architectural concepts were molded to an established framework of indigenous institutions and building traditions, creating a distinctive and unique urban form. These coastal cities were shaped by the nature of indigenous economic, social and political institutions which formed the underlying features of Swahili urban culture. The addition of new building forms and styles were at times influenced by foreign prototypes and designs, yet they were also conditioned by indigenous building traditions, and thus often resembled only superficially the models that inspired them. The result was reinterpretation through a selective and adaptive process of design, leading to the creation of a coherent architectural language and urban landscape which accommodated the needs and evolving institutions of late medieval Swahili society.

This hybridization process was a form of "creolization," which may be characterized as the "creative interplay," resulting from sustained cross-cultural interaction between unequal economic partners — as in a core and periphery relationship — and resulting in a new synthesis of cultural expression.[41] This process was an inventive act, whereby interacting groups formed a "new construct" of society, one that was not a corrupted image of the dominant culture, but rather a hybrid which absorbed newcomers by inclusion through the acceptance of common cultural practices.[42]

The creation of the medieval Swahili city resulted from the interaction between imported ideas and established African institutions and building practices. Their distinctive architectural forms and urban traditions contributed to the evolving cultural identity of medieval Swahili society, which allowed Swahili communities to create and maintain multiple affiliations and associations to further their own economic and cultural objectives. Their cities identified the Swahili people as economically part of the cosmopolitan culture of the Indian Ocean basin, linked by seafaring trade and commerce; this was ideologically a part of the greater Muslim world, to which they turned for inspiration and direction, yet ultimately residents of the African continent, which formed the basis of their social institutions and building traditions. Such a identity differentiated them both from the societies of the African interior and from those societies across the sea with which they maintained important cultural and commercial linkages.

NOTES

1. R.F. Burton, *Zanzibar: City, Island, and Coast*, vol.2, 1872, p.39.

2. J. Aldrick, "The Nineteenth-Century Carved Wooden Doors of the East African Coast," *Azania*, vol.XXV, 1990, pp.12-14.

3. M. Sabini, *The Architecture of Mombasa*, vol.2, privately reproduced study, 1993, pp.223-224.

4. Aldrick, "Nineteenth-Century Doors."

5. G.M. Theal, *Records of South-Eastern Africa*, vol.VI, London, William Clowes and Sons, Ltd., 1898, 1899, and 1901; facsimile reprint, Cape Town, C. Struik, Pty., Ltd., 1964, p.179.

6. Ibid., p.236.

7. Ibid., p.255.

8. Theal, *Records of South-Eastern Africa*, vol.III, p.93.

9. T.R. Gensheimer, "At the Boundaries of Dar-al-Islam: Cities of the East African Coast in the Late Middle Ages," Ph.D. diss., University of California, Berkeley, 1997, chapter 2.

10. D. Nurse and T. Spear, *The Swahili: Reconstructing the History and Language of an African Society 800-1500*, Philadelphia, University of Pennsylvania Press, 1984; M. Horton, "The Settlement of the Northern Swahili Coast," Ph.D. diss., University of Cambridge, 1984; and G. Abungu, "Communities on the River Tana, Kenya: An Archeological Study of Relations between the Delta and the River Basin, 700-1890 A.D.," Ph.D. diss., University of Cambridge, 1989.

11. J.C. Wilkinson, "Oman and East Africa: New Light on Early Kilwan History from the Omani Sources," *International Journal of African Historical Studies*, vol.14, no.2, 1981, p.279; G.F. Hourani, *Arab Seafaring in the Indian Ocean in Ancient and Early Medieval Times*, Princeton, Princeton University Press, 1951, p.79; Sir A. Stein, *Archaeological Reconnaissances in North-Western India and South-Eastern Iran*, London, MacMillian and Co., 1937, p.202; and David Whitehouse, "Siraf: A Medieval City on the Persian Gulf," *Storia della Citta*, no.1, 1976, p.42.

12. M. Horton, "The Swahili Corridor," *Scientific American*, vol.257, no.3, September 1987, pp.86-87.

13. K.N. Chaudhuri, *Trade and Civilization in the Indian Ocean*, New York, Cambridge University Press, 1985, pp.103-106.

14. J.L. Abu-Lughod, *Before European Hegemony: The World System, A.D. 1250-1350*, New York, Oxford University Press, 1989, p.271.

15. In the early tenth century al-Mas'udi mentions that "the sea of Zanj ends with the land of Sofala and Waqwaq, which produces gold and many wonderful things." G.S.P. Freemen-Grenville, *The East African Coast: Selected Documents from the First to the Earlier Nineteenth Century*, London, Rex Collins, 1975, p.15.

16. B.G. Martin, "Arab Migrations to East Africa in Medieval Times," *The International Journal of African Historical Studies*, vol.VII, 1975, pp.370-371.

17. Ibid., pp.378-380.

18. C. Meyer, *Glass from Quseir al-Qadim and the Indian Ocean Trade*, Chicago, The Oriental Institute of the University of Chicago, 1992, p.101; and E.A. Alpers, "Gujarat and the Trade of East Africa, c.1500-1800," *The International Journal of African Historical Studies*, vol.IX, 1976, pp.24-26.

19. T. Raychaudhuri and I. Habib, *Cambridge Economic History of India*, vol.1, New York, Cambridge University Press, 1982, p.150.

20. Freemen-Grenville, *The East African Coast*, p.126.

21. C.M. Kusimba, "The Archaeology and Ethnography of Iron Metallurgy on the Kenya Coast," Ph.D. diss., Bryn Mawr College, 1993, pp.65-66; and Horton, "Northern Swahili Coast," p.264.

22. Horton, "Northern Swahili Coast," p.328; and "The Swahili Corridor," p.91.

23. J. de V. Allen, "Swahili Culture and the Nature of East Coast Settlement," *International Journal of African Historical Studies*, vol.XIV, 1981, p.319.

24. A.M. Shieriff, "Trade and Underdevelopment: The Role of International Trade in the Economic History of the East African Coast before the 16th Century," *Hadith*, vol.5, 1976, p.16.

25. U. Ghaidan, *Lamu: A Study of the Swahili Town*, Nairobi, Kenya Literature Bureau, 1975, pp.35-36; J. de V. Allen, *Lamu Town: A Guide*, Mombasa, Rodwell Press, Ltd., 1986, pp.13-16; and A.H.M. el Zein, *Sacred Meadows: A Structural Analysis of Religious Symbolism in an East African Town*, Northwestern University Press, 1974, pp.11-13.

26. Freemen-Grenville, *The East African Coast*, pp.30-31.

27. Theal, *Records of South-Eastern Africa*, vol.I, pp.66-67; and A. Da S. Rego and T.W. Baxter, *Documents on the Portuguese in Mozambique and Central Africa*, vol.1, Lisbon, National Archives of Rhodesia and Nyasaland, 1962, pp.397-399.

28. Gensheimer, "At the Boundaries of Dar-al-Islam," pp.31-41.

29. P.S. Garlake, *The Early Islamic Architecture of the East African Coast*, Memoir no.1 of the British Institute of History and Archaeology in East Africa, London, Oxford University Press, 1966, pp.88-89.

30. J. Kirkman, *Gedi: The Palace*, The Hague, Mouton & Co., 1963, pp.18, 22.

31. Rego and Baxter, *Documents on the Portuguese*, p.531.

32. M. Horton, "Early Muslim Trading Settlements on the East African Coast: New Evidence from Shanga," *Antiquaries Journal*, vol.67, 1987, p.312; and D. Whitehouse, *Siraf III: The Congregational Mosque and Other Mosques from the Ninth to the Twelfth Centuries*, London, The British Institute of Persian Studies, 1980, pp.30-57.

33. Horton, "Northern Swahili Coast," pp.184-185.

34. A.H.J. Prins, *The Coastal Tribes of the North-Eastern Bantu*, Ethnographic Survey of Africa, London, International African Institute, 1952, pp.87-90.

35. R. Lewcock, "Zanj, The East African Coast," in P. Oliver (ed.), *Shelter in Africa*, London, Barrie and Jenkins, 1971, p.89; and "Architectural Connections between Africa and Parts of the Indian Ocean Littoral," *Art and Archaeology Research Papers*, vol.9, 1976, p.18.

36. Freemen-Grenville, *The East African Coast*, pp.27-28.

37. Shieriff, "Trade and Underdevelopment."

38. Abu-Lughod, *Before European Hegemony*.

39. J.N. Pieterse, "Globalization as Hybridization," in M. Featherstone, S. Lash and R. Robertson (eds.), *Global Modernities*, London, Sage Publications, 1995, p.45.

40. Ibid., p.62.

41. U. Hannerz, *Cultural Complexity: Studies in the Social Organization of Meaning*, New York, Columbia University Press, 1992, pp.264-265.

42. E. Brathwaite, *The Development of Creole Society in Jamaica, 1700-1820*, Oxford, Clarendon Press, 1971, pp.296-297.

2

Orchestrating Difference, Performing Identity: Urban Space and Public Rituals in Nineteenth-Century Izmir

Sibel Zandi-Sayek

On May 27, 1842, *l'Écho de l'Orient*, the French newspaper of Izmir (Smyrna), ran a long report describing the unusual pomp of the Corpus Christi procession. For the first time, the celebration had emerged from the precinct of the Church of St. Polycarp — the oldest major Catholic Church in town — onto the streets of Izmir. The areas marked on the itinerary of the procession had taken a festive look the night before. Façades of buildings had been decorated with myrtle and flowers, and the crowd waited anxiously for the solemn ceremony, which was unprecedented in the city. When the procession began, a Turkish military guard escorted the holy sacrament through the streets, obliging onlookers to respect its passage. At the center of the procession was the archbishop, holding the Eucharist under a grand canopy. He was preceded by the clergy and students of the city's Catholic schools, followed by the French consul — "the safeguard of the Catholics of the Orient"[1] — and by many of the consuls of the other Catholic countries of Europe.

The Corpus Christi procession — the carrying of the Eucharist across town — was a medieval Catholic ceremony which first appeared in the context of mid-thirteenth century Europe. The part of the rite which did not have to be enacted within a church was performed out of doors as a public event.[2] During the nineteenth century, the devotion was still alive throughout the Catholic world, but was frequently under attack because anti religious feelings among European ruling elites saw the ceremony as a danger to advancement. Ironically for *l'Écho de l'Orient*, whose readership consisted mostly of members of foreign colonies and the French-speaking local elite, the celebration signaled an important step forward. It marked the presence of a level of freedom and civility which was lacking even in Europe. Remarking that elsewhere such events were constantly hindered by disbelievers, the reporter proudly recounted the Corpus Christi procession in Izmir to his coreligionists as an indication of increased tolerance in Ottoman territories:

> Today, religious animosity has disappeared among us. The Catholic population of Smyrna saw with satisfaction Greek merchants decorate . . . with hangings and flowers their shops located on the processional route. All took place in the most perfect order and we cannot sufficiently acknowledge the concern of His Excellence Sabri Pasha on this solemn occasion.[3]

The parade was performed along Frank Street, the most prominent thoroughfare of Izmir, and before a diverse population of Catholics, Greeks, Armenians, Jews and Muslims. Its visibility and its diverse audience suggested that it represented more than an incidental display of religious tolerance. The presence of a Turkish guard of honor at the disposal of the archbishop, the participation of the French consul as representative of the "Catholic nations," and the pride in having an all-inclusive audience further revealed the effort by the Catholic community to assert itself in relation to other groups in the city.

The emergence of the Corpus Christi procession corresponded to a critical period in the history of non-Muslim Ottoman communities. The Gülhane edict of 1839 and the imperial rescript of the *Hatt-ı Humayun* in 1856, which constituted the legal basis for a series of reforms known as *Tanzimat*, sought to redefine the relation between local religious groups and Ottoman authorities. These decrees called into question the nature and terms of plural coexistence in the empire. In the prereform period, the Ottoman constitution had organized the heterogeneity of the populations on a confessional basis. Armenians, Greeks, Jews and Muslims had been considered separate nations, called *millet* and controlled by their own religious hierarchies and internal organizations.[4] Yet even though Christians and Jews had enjoyed a certain degree of autonomy within their nations and had been allowed to practice their religions, they had still been subject to various discriminatory rules which had legally placed them in a second-class position.[5] These restrictions had applied to various spheres of life, ranging from judicial, political and economic matters to purely ecclesiastical ones such as not being allowed to build new places of worship, hang or ring church bells, or carry crosses in public.

In an attempt to end the aspirations of non-Muslim communities for independence, the reform decrees abolished such constitutional distinctions. They promoted a new political community based on the idea of uniform citizenship. This represented a new way of thinking, elevating non-Muslim individuals to the status of equal citizen rather than considering them members of autonomous, yet subject, communities. It also presented reformers with the challenging task of introducing a unified Ottoman allegiance to override the heterogeneous social constitutions. Granting full equality to all Ottoman subjects did not eliminate, however, the propensity of local communities to express their religious and national differences. To the contrary, it provided these communities with an opportunity to enact these differences more publicly. As groups gained more freedom, they were not hesitant to practice their particular rituals in public. In several towns, church bells were rung, crosses were carried in procession, and wine shops were opened in marketplaces. The Corpus

Christi celebration was simply one instance of the more general freedom to carry communal differences into the public realm. Izmir, the second largest city on the Anatolian peninsula and one of the most prominent port cities of the empire, was a venue for a wide range of public ceremonies. In the decades following the 1840s, Orthodox Easter, various saints' days, the sultan's accession to the throne, and the coronation of foreign emperors were all important public events celebrated with similar pomp. These events overlapped in space but were disparate in meaning, and they juxtaposed divergent political ambitions — the Ottoman authorities' effort to create a unified political allegiance and local groups' predilection to affirm their national and religious differences.

In this chapter, I discuss, through description and analysis of a range of public performances, how these divergent ambitions were played out in the urban landscape of Izmir. I contend that in the face of the Tanzimat freedoms and constraints, public celebrations offered a critical tool for local communities and Ottoman authorities to position themselves in relation to one another. Although civic and religious rituals have recently provided an important site to examine local political action, conflict and consensus in the context of nineteenth and twentieth century American and European cities, the role of urban space in the performance of ethnic and national identities has been largely neglected.[6] This chapter is an attempt to begin to rectify that neglect by situating public festivities in both the social and the physical context of Izmir. By analyzing the form, location and meanings of these events, I seek to demonstrate how the spatiality of these performances was critical for mediating competing social and political claims. More specifically I explore how the physical and social spaces of Izmir were reinvented, remade and differentiated through the use of special decorations applied to the façades of buildings and the gathering of crowds as participants and spectators. All along the route of these events, the presence of costumes, music, and other visual and auditory effects — indeed, the very movement between designated locations — had a symbolic resonance that distinguished these areas from their everyday use. Understanding this range of spatial strategies is not only crucial for grasping the local specificities of Tanzimat Izmir, but also the very nature of negotiating emerging political and ideological positions through spatial practices.

THE SOCIAL AND SPATIAL LANDSCAPE OF A NINETEENTH-CENTURY OTTOMAN PORT CITY

Located on the eastern coast of the Aegean Sea and endowed with a secure, deep-water anchorage at the junction of several trade routes, Izmir was an important destination for long-distance caravans from Anatolia. According to Ottoman registers in the 1580s, the city consisted of eight Muslim and one Orthodox Greek neighborhood, and counted about 2,000 inhabitants. By 1650 its population had soared to 30,000-40,000, drawing particularly Armenians, Greeks and Jews. During these decades Dutch, English, French and Venetian merchants were also attracted to this growing trade post, and established consular representation.[7] The Ottoman historian Daniel Goffman has explained this growth as the result of the confluence of internal events and structural changes in world trade. In the early decades of the seventeenth century, rural rebellions in Anatolia — known as the *celali* rebellions — created social distress in the countryside, forcing the dispossessed to seek refuge in protected towns like Izmir. At the same time, changing world trade routes made Izmir a strategic location for immigrants in search of economic opportunity. Armenians settled mostly by the end of the seventeenth century, principally because of the silk trade with Persia which linked Izmir to an already extensive trade system stretching from China to Europe. Jews migrated to the city in the sixteenth century from Iberia and Italy to serve mostly as translators, customs officials, and tax farmers. During this period, Greeks — mostly from the island of Chios — already boasted a small settlement in the city, and in the following centuries, with continual migration from the Aegean islands, they far outgrew other communities.

Until the demise of the Ottoman Empire and the Turkish war of independence of 1922, Izmir was a multireligious, multinational and multilingual center. As in Beirut, Alexandria, Salonica, and other principal Ottoman port cities, local Armenians, Jews, Muslims and Orthodox Greeks lived alongside colonies of foreign merchants. The physical landscape of the city expressed and sustained this social plurality. Izmir stretched north-south on a flat strip of land abutting the water to the west, and extended up into the hills bounding this flatland to the southeast. From the bay, the city presented a fairly uniform and continuous physical fabric running along a two-and-a-half-mile shoreline and over the acclivity of Mount Pagus (or Kadifekale), which was crowned by a thirteenth century Genoese castle. A dense commercial zone — the bazaar — adjoined the water, and to the east of this commercial zone were residential neighborhoods. Typical nineteenth century plans of the city differentiated the Turkish, Armenian, Greek, Jewish and Frank quarters, allocating distinct parts of town to different nations **(Figure 2.1)**. Thus, the majority of the Turkish population lived on the slopes of the hills. Jewish areas were located adjacent to these high quarters, extending down onto the plain. Next to them was the low-lying Armenian quarter adjoining the Caravan Bridge Road — the main entry point to the city from the interior until two railways started serving the city in the 1860s. The Greek neighborhoods spread to the north on the rest of the city's low-lying areas, leaving between them and the shore a linear zone known as the Frank quarter.

The residential layout of Izmir followed in broad lines an organization defined by religious and national differences. Localized networks of institutions such as mosques, churches or synagogues reinforced a spatial sense of group identity. The geography of quarters, however, neither confined nor totally defined the lives of inhabitants. Individuals associated with various cultural forms combined religious and secular practices to deal with the circumstances that stirred their social world.

In the early nineteenth century, the Tanzimat reforms attempted to impose a more homogeneous social organization based on political allegiance, and thus alter

Figure 2.1. Izmir in the 1880s. (D. Georgiades, *Smyrne et l'Asie Mineure du point de vue économique et commercial*, Paris, Chaix, 1885.)

the terms of this plural existence. The reforms were based on the assumption that promoting equality among the empire's subjects would generate firmer state loyalty and unite peoples "by the cordial ties of patriotism."[8] Achieving this ideal, however, was tied to how these changes were perceived, contested and reworked by the subject peoples. Thus, the reforms had particular inscriptions and implications in the socially mixed context of Izmir. The heterogeneity of Izmir was not limited to the diversity of its Ottoman residents. There was also a numerically small, but economically influential, colony of foreigners who, concurrently with the early Tanzimat, began to take a more active part in the politics and administration of the city.[9] Until the late eighteenth century, the foreign merchants and colonies established in Ottoman port cities had been bound both by local codes and the strict rules of their respective nations regarding the duration of their stay, the location of their residence, and the pattern of their relations with the local people. But in the early nineteenth century, as limitations imposed by their homelands regarding their settlement in the city or intermarrying with locals were relaxed, foreigners residing in Izmir had begun to enjoy more personal freedom. However, increased freedom often interfered with the workings of local Ottoman administration. Charles Wood, a member of the British colony residing in Izmir, wrote in 1866 that the position of the governor-general (*vali*) of Izmir was peculiar and difficult. According to Wood:

> The city itself stands apart from almost every provincial capital in the circumstance of its being inhabited by a population of which the predominant element is Christian, and the distinguishing characteristic a very independent demeanor towards the local authority. . . . [I]t has ever been a difficulty with the governors of Smyrna to conciliate the impatient requirements of the Christian part of the population . . . with the habits and notions of the Mussulman part.[10]

The frequent changes in the governor's office attested to this fact. Between 1872 and 1875, ten *valis* were appointed to Izmir, hinting at increased surveillance and distrust by the central authorities.[11] Unlike the capital, Istanbul, where the Muslim ruling elite and the central government always remained dominant, in Izmir (as in other provincial cities), power and authority was diffused more widely among local groups. Thus, non-Muslim communities of Greek Orthodox, Armenian, Jewish and Roman Catholic subjects had more leeway to negotiate their position within an ever-changing social and political situation.

Within such a context, the decision in 1842 by Bishop Moussabini to carry the Corpus Christi procession to the streets of Izmir for the first time was not an incidental act. In Ottoman cities the Catholic population had come to constitute a nation (called the *Latin* or *Katolik milleti*) much later than had the Greek Orthodox, Armenian and Jewish populations. Only by the early nineteenth century had the Catholic population of the Ottoman Empire established delegates in major cities such as Istanbul, Izmir or Chios. At the time, they frequently sought the protection of the French government and attended Catholic missionary schools, particularly the Lazarist institutions established throughout the eastern Mediterranean.[12] Then, in 1841, the French foreign ministry allocated a substantial budget to be spent on Ottoman Catholics and on their religious institutions — a portion of which was sent to the Izmir delegate.[13] Such French support, coupled with the new political climate of the Tanzimat, encouraged the numerically small but politically strengthening Catholic community of Izmir to affirm their presence overtly. The choice of Frank Street — the main traffic artery — as a parade route was not random: it was the most public area in the commercial part of the city.

FRANK STREET — AN EPITOME OF CULTURAL PLURALITY

In Izmir, social and cultural interaction was most explicit in and around Frank Street.[14] The street began at the northern edge of the bazaar and contained, as its name implies, most of the city's principal European establishments **(Figure 2.2)**. Many foreign consulates and European merchant houses were located within the area bounded by Frank Street and the shoreline. Until the late

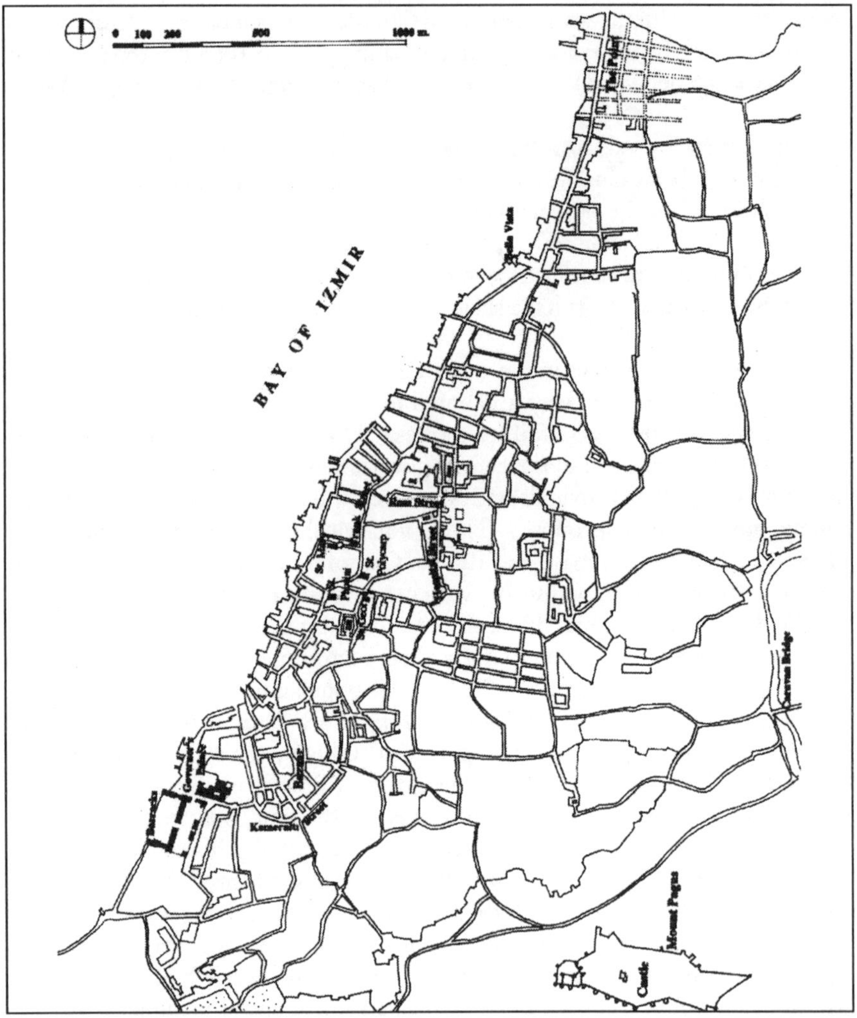

Figure 2.2. Izmir in the 1850s. (Drawn by author based on 1854-1856 map of Izmir by Luigi Storari, and on map of Izmir in *Murray's Handbook for Travellers in Turkey in Asia*, London, John Murray, 1878.)

eighteenth century, Frank Street and its vicinity had been occupied mainly by merchants of European descent. However, by the early nineteenth century, as regulations were abolished that constrained the location and type of residences that could be occupied by Europeans and non-Muslim Ottomans, Europeans, as had Armenians and Greeks, scattered more widely across the city. Thus, by mid-century there was no Frank quarter proper, and, in fact, no exact separation of quarters along ethnic/religious lines. Although the area kept its name, its offices and stores were equally owned by Greeks and Europeans, and the residential area stretching southeast from Frank Street was occupied by a mixed group.[15]

Frank Street epitomized in many ways the diversity of Izmir, and it lent itself to multiple representations. In descriptions of the city, Frank Street always occupied a central place. For some travelers it was the center of European and Christian neighborhoods; for others it was a meeting point of races and languages. Romaic Greek and dialects of French and Italian were heard at all times of the day. Frank Street was also a bustling marketplace where goods from all over the world were on display. "World's trading posts met on Frank Street," wrote Charles Reynaud on his visit in 1844.[16] Yet even though the street epitomized Izmir's connection with the larger network of goods and commodities, it was neither a grand nor a wide thoroughfare. In the 1850s it was a narrow, two-lane street located about 200 meters from the sea, its contour following the irregular sixteenth century historical shoreline. Most characteristically its western edge opened to the shore through a series of frequent, narrow, privately owned vaulted passageways flanked by shops. These properties, called *verhanes*, often also had wharves that extended out to sea to allow merchandise to be shipped directly to and from them.[17] One result was that ships often smuggled their goods to the doors of the verhanes without paying custom dues.

Frank Street remained an important artery throughout the nineteenth century, even though in later years it lost its relative proximity and direct connection to the shore. Beginning in 1869 a wide quay was developed on land reclaimed from the sea along the entire length of the city (starting at the military barracks to the south and extending to the Point area to the north). The new quay added several regular urban blocks and two spacious boulevards — the first and the second *cordon* — running parallel to the shore. This linear land addition adjoined the Frank quarter, expanded the commercial areas, and increased to 400 meters the distance from Frank Street to the shore **(Figure 2.3)**. Their new distance from the shore meant the verhanes stopped being used for their old functions, and were occupied by a variety of wholesale and retail shops. In 1875 after the quay was complete, the waterfront area drew new businesses, hotels, restaurants, cafes, and other entertainment functions. But Frank Street continued to house the city's principal retail stores, such as the large dry-goods stores Au Bon Marché and Orosdi-Back, as well as several hat-makers, goldsmiths and pharmacies. Two Protestant Chapels, three Catholic Churches, and, further down, a Greek Orthodox Cathedral, were also to be found along Frank Street.

In a novel based on his personal experience of late nineteenth-century Izmir, Kozmas Politis gave powerful insight into the everyday life of a working-class

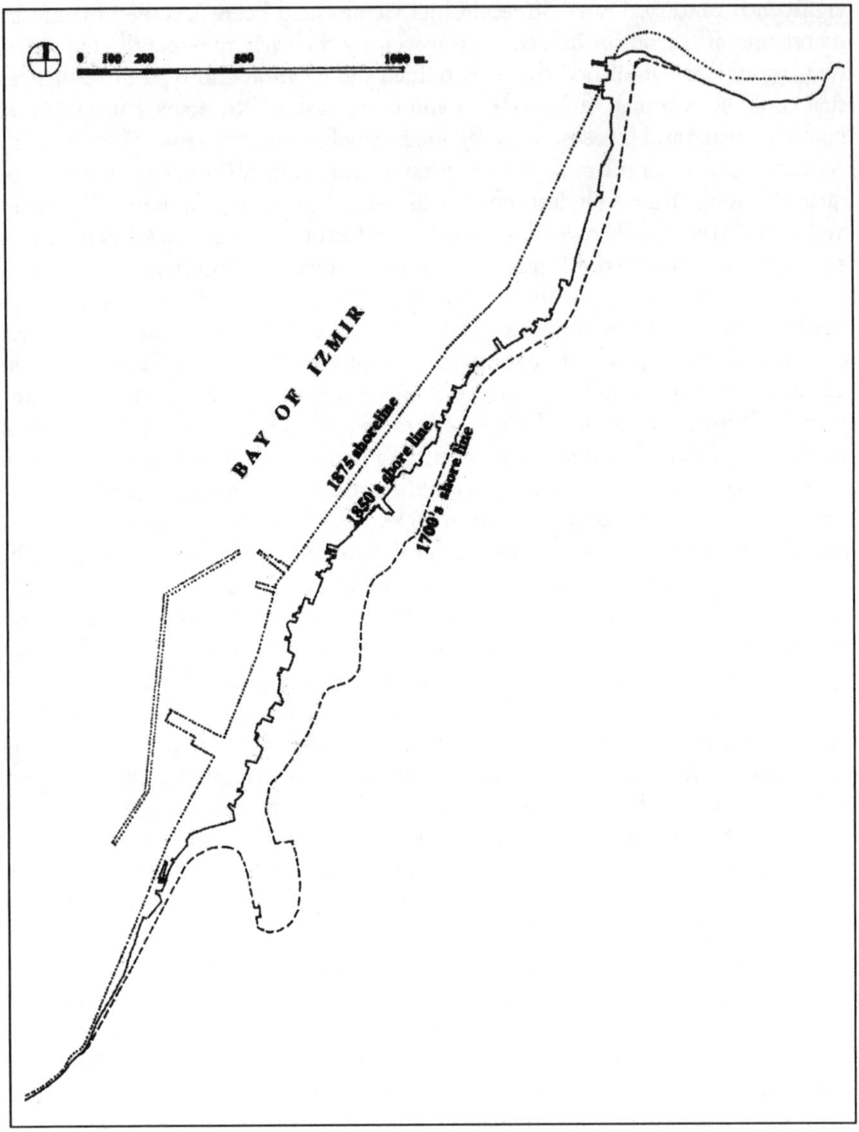

Figure 2.3. Changes in the shoreline of Izmir. (Drawn by author.)

Greek neighborhood in the city. "When 12 o'clock rang on the bell of the major Greek Cathedral, St. Photini, Frank Street began to thrive," remarked Pandelis, the young Greek protagonist in Politis' novel.

> Women would crowd the street and the stores carrying all sorts of goods. . . . The clerks working at the banks would take their lunch break, gentle-

men would leave their offices and people would exchange greetings in all languages:
— Kalimera!
— İyi sabahlar hanımefendı
— Bonjour mademoiselle
— Hello Jim!
as they did at Sunday noon as they walked along the quay.[18]

However, not only the genteel appeared on Frank Street. All classes of people had to walk through the area to get to work. As a link between residential neighborhoods and the bazaar, the part of Frank Street stretching from Fasula Place to the bazaar was the busiest and most animated part of the entire city, and two-thirds of the city's retail shops, stores, and offices were located there. In Politis' novel, Pandelis loved watching the grocers, butchers, and fish sellers; and he admired the horses lined up at the omnibus station in front of the great *khan* on Fasula Place. He also took pride in the location of the Greek pharmacy at Fasula adjacent to the French one. Its window had the symbol of the Greek Kingdom and the inscription "wholesale with the permission of His Excellency the Greek King." Local newspapers, however, continually drew the attention of the public administration to the "sad state of Fasula," where clusters of coffee shops, butchers, grocers, and fish sellers' stalls blocked movement through the area at all times. Indeed, the fish market on Fasula was removed several times, but commercial encroachment on these spaces was never eliminated. At all times, meat, vegetables, poultry, and all sorts of goods were displayed for sale, creating an offensive sight for the genteel. "We hope that his Excellency [the governor] would rid us from the stalls which prevent circulation. Fishsellers stand in front of the door of the Lazarist church and poison the Frank quarter with stoves placed in the middle of the street risking any moment to cause an accident," lamented a reporter for the *Journal de Constantinople*.[19]

Past Fasula Place toward the bazaar a number of institutions fronted onto Frank Street. To the right was the English consulate and chapel, the U. S. consulate, the verhane of the Levantine Club, the verhane of the Imperial Ottoman Bank, and the Catholic Church of St. Mary **(Figure 2.4)**. Further along to the left were the Catholic Church of St. Polycarp and the archbishop's house, and a few steps further the Greek Orthodox Cathedral of St. Photini. A short distance from there the street ended in the cloth market, the entrance to a dense web of commercial streets that constituted the Turkish bazaar. At Fasula Place, Frank Street extended northwest (its name changed to Trassa Street) toward Bella Vista corner — a five-way intersection marked by the French hospital and a Turkish guardhouse; before the new quay, Bella Vista abutted the shore, and in this part of town boats could be hired for excursions on the bay. Numerous coffee shops also clustered on pilotis, with terraces extending out to the sea. These places served coffee, drinks and ice cream, had European newspapers, and some evenings held comedies and musical performances. The Bella Vista area was the favorite promenading place for the upper classes, local and foreign alike.

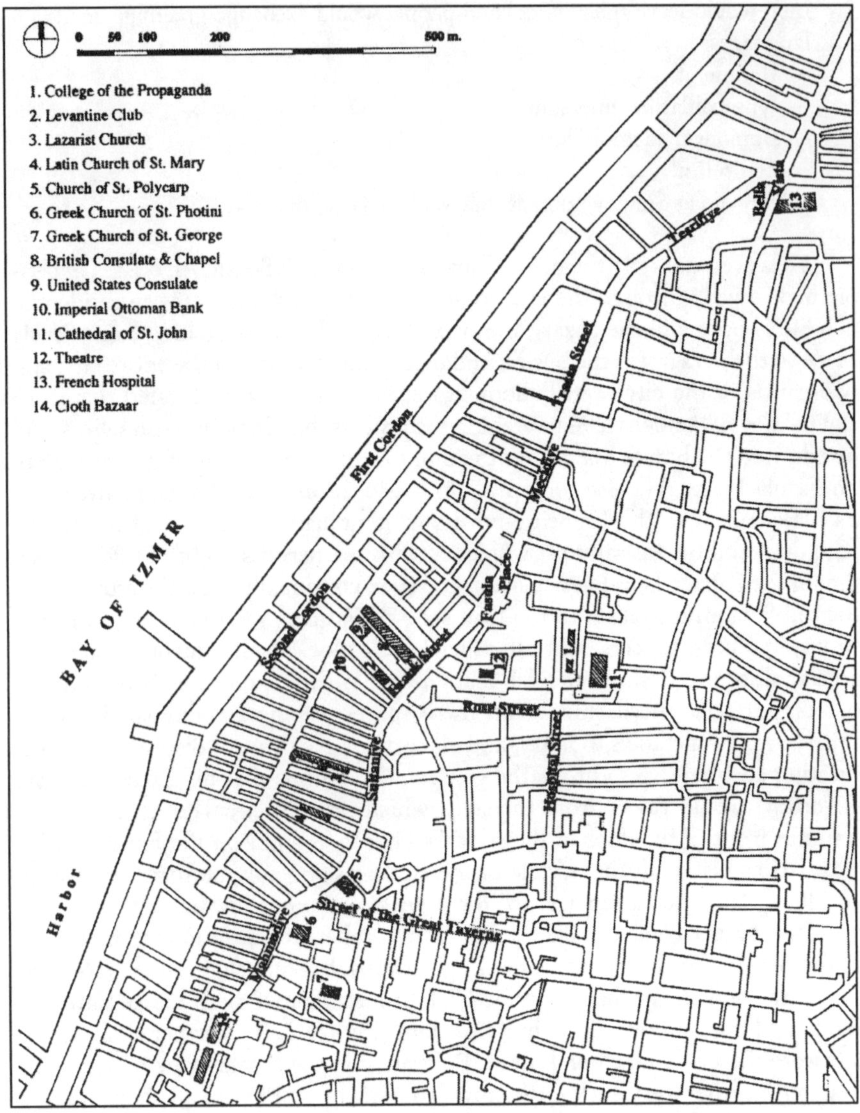

Figure 2.4. Frank Street and vicinity in the 1880s. (Drawn by author based on 1876 Izmir map by Lamec Saad, and on Karl Baedeker's map of Izmir in *Konstantinopel und der Westliche Kleinasie*, Leipzig: Verlag von Karl Baedeker, 1905.)

Perpendicular streets connected Frank Street to the inner (eastern) parts of town. Opposite the English consulate was a street leading to Teatro Cammarano and the French College of the Propaganda to the left, and into Hospital Street and upper Rose Street to the right. Rose Street was mostly residential, and was renown for its beautiful mansions, occupied by local Greeks. The street drew the attention of many visitors for "the gracious spectacle offered by the sight of so many beautiful women dressed in white."[20] On regular days, sailors, travelers and dandies wandered at all hours to see the young Greek women sitting at their doorsteps. Also opening onto Frank Street and adjoining St. Photini was the Street of the Great Taverns. This street was famed for its numerous coffee shops and taverns, which offered a much different atmosphere than the coffee shops on Bella Vista. These places attracted Europeans, Greeks and Turks of relatively modest conditions, as well as sailors looking for a quick meal or a coffee. Particularly in the evening, these taverns were transformed into exclusively male spaces where alcohol abounded and frequent brawls took place.

Multiple social lives met in this part of town. Class, religious and linguistic differences coexisted in and around Frank Street. In the spaces of the street the social and cultural entanglement was rendered visible by the different ways people related to the public space. Various names were even used to designate Frank Street itself. Some called it Fasula, or Trassa, and in official documents different segments of it were named after members of the Ottoman dynasty: Mahmudiye, Sultaniye, Mecidiye, and Tesrifiye. In this way the Ottoman authorities also acknowledged the importance and centrality of the street to the city. Nineteenth century maps of Izmir included these various appellations simultaneously, not only exhibiting the equivocality of this part of the city, but also its contested character.

PUBLIC CEREMONIES AS ENACTMENT OF DIFFERENCE

The Corpus Christi procession provides dramaturgical keys and cues for understanding how emerging communal allegiances and social hierarchies were constituted and reinforced through public performances. The event took place in and around Frank Street, and the movement between stations was carefully traced to produce an exclusively Catholic effect. Led by Bishop Moussabini, head of the Catholic Church in Izmir, the march began at the College of the Propaganda, a school built in 1839 by the Lazarists.[21] In the early morning several squads of Turkish soldiers had been stationed at intervals along Frank and Rose Streets, and Salih Pasha, the governor himself, had crossed the area several times to herald good order and tranquillity. Participants were strictly positioned within the procession. Guards (*cavass*) of the governor and of the consulates opened the march. They were followed by twelve members of each brotherhood, all carrying banners and torches.[22] Next came the students of the College of the Propaganda (the Lazarists), accompanied by the friars of the Christian schools. They were followed by 200 young girls of the Sisters of Charity, dressed in white and holding banners representing the Virgin Mary. Scattered among them were the girls of St. Vincent de Paul, followed by a choir of artists chanting religious hymns to the sound of various instruments. Official translators (*dragomans*) of the Catholic countries' consulates preceded the clergy, and twenty altar boys strewing flowers and incense cleared the way for the archbishop, who carried the Corpus Christi under a grand canopy. The French consul and the consuls of other Catholic nations immediately followed, while the Turkish guard of honor, assigned by Governor Salih Pasha, closed the procession and separated it from the crowd.

Leaving the College, the procession walked to upper Rose Street, which had been decorated for the event, and continued down this street to reach Frank Street. On Frank Street the canopy stopped in front of an altar placed at the entrance of the Levantine Club, next to the British consulate. The altar had been built by the Sisters of Charity, with the assistance of club members, and had been approved by the archbishop. According to the report in *l'Écho de l'Orient*, at the altar the archbishop blessed the crowd, which was composed of "Catholics, Greeks, Muslims, Armenians, etc., wordlessly attending the performance." The march then continued along Frank Street, which was lined with shops owned mostly by Greek merchants and covered with hangings and flowers. The procession next went to the Lazarist Church, continued a few steps down to the establishment of the Sisters of Charity, went on to the Church of St. Mary, and finally ended at the Church of St. Polycarp. Accompanying the blessing at each of these station points, the French corvette *la Cornaline*, an Austrian schooner, and a Tuscan three-master anchored in the harbor fired 21-gun salutes.[23]

The itinerary of the Corpus Christi procession demarcated the territory of the Frank quarter as an exclusively Catholic domain. By marching through Frank and Rose Streets, the two main arteries along which the city's Catholic institutions were located, it traced a symbolic sphere of influence, claiming this socially mixed area as an exclusively Catholic space. The main actors — the College of the Propaganda, the Lazarists, and the Sisters of Charity — also portrayed a predominantly Catholic presence in town. Although these institutions were an intimate part of the everyday social and educational life of Izmir, functioning generally in a more ecumenical way, for the event they were recast as places of ultimate Catholicism. In actuality, the college taught French and Greek and not only catered to Catholics but to upper-class families, including Greeks, Armenians, and occasionally Muslims. This was a source of contention among the authorities in Istanbul, who disapproved of having Muslim children educated in Christian institutions.[24] Similarly, the Sisters of Charity were regularly praised in local papers for their benevolent work, which aimed at helping people regardless of class or religious difference.

The procession attempted to link these institutions physically and symbolically. To create a seamless Catholic space, the itinerary sewed together selective parts of the town. The processional plays performed at the stations along Rose and Frank Streets connected the College of the Propaganda with the Church of St. Mary, the Church of St. Polycarp, or the institution of the Sisters of Charity — all Catholic institutions, though supported by different nations (mainly France, Austria and Italy). Although most of the stops on the route consisted of religious institutions, they were not exclusively so. Curiously, the itinerary included a secular institution, the Levantine Club — also known as the

European Casino. Izmir had several such casinos, or family clubs. The Levantine Club, like the Greek one, was supported by the subscription of its members and had, among other facilities, a news room, a card room, a billiard room and a ballroom. By erecting an altar in front of it, Catholics appropriated the club into the religious idiom, thereby suggesting an idea of Catholicism that was compatible with modern and secular institutions. At the same time, even though the procession stopped at the Levantine Circle adjacent to the English consulate and its chapel, the consulate was not mentioned as part of the route. Thus, if some places were reinvented as Catholic, others were clearly edited out to bring the itinerary into conformity with Catholic ideas.

On such processional occasions, not only was the route carefully selected, but also it was prepared days in advance. Walls of buildings along the way were covered with tapestries, cloths and banners, and decorated with pictures, foliage, lanterns and small statues. Altars or statues of the Virgin Mary, adorned with flowers and lights, were placed at street corners, and cloth was stretched between houses to provide shade for the crowd. In addition, care was taken to coordinate the salutes fired by the warships in the harbor with the arrival of the procession at each of the principal station points. Such announcements from the Austrian, Italian and French ships symbolically unified these nations under the banner of Catholicism. Meanwhile the chanted hymns and smell of incense spread during the procession gave a church-like quality to the space of the street. Such visual and auditory effects opened up a different way of experiencing everyday urban spaces. They inscribed a distinctive spatiality onto Frank Street, raising temporary claims of authority over it.

At one level, the Corpus Christi procession provided a tool for local Catholics to position themselves vis-à-vis the Ottoman authorities and the French nation. Catholic missionary activity had always received official French support, and the ceremony was a matter of delicacy because it defined a double allegiance to competing sovereignties. On one side was the French consul, walking in the footsteps of the archbishop, representing the French nation and the other Catholic nations of Europe. On the other was the watchful gaze of Ottoman authorities. The parade organized and mediated this dual allegiance. The ambivalent position of the consul as a civic and religious leader was important to this balance, because it diffused any direct threat toward the Ottoman authorities. Ottoman support was indeed secured throughout the performance, and the governor's involvement and the presence of a Turkish squad confirmed and reinforced the position of local Catholics as equal citizens of the reformed Ottoman Empire, free to practice their religion.

At another level, the Corpus Christi procession allowed Catholics to position themselves in relation to other communities in the city — particularly Orthodox Greeks, who constituted the majority of its Christian population. Reports on the procession made continual reference to the great enthusiasm shown by Orthodox Greeks. Whether fueled by a general euphoria over the Ottoman authorities' letting such a performance happen in public, or by its novelty, Greek shopowners competed in decorating their windows for the event. These allusions to Greek cooperation meant more than simple Christian comradeship, and must be seen against the background of complex intercommunal relations in mid-nineteenth century Izmir.[25] A feeling of dislike and distrust existed between the Greek Orthodox and Roman Catholic communities, fostered in particular by the attitudes of the priesthoods of both persuasions, and these feelings were often articulated in public. In 1869, for example, at the funeral of the Greek archbishop (which also took place along Frank Street), while Ottoman and foreign officers of various religions followed in uniform (and the local Armenian clergy, Protestant pastor, and Jewish delegates were also present), the Catholic priests who encountered the procession "simply watched it pass without even taking off their head gear."[26] Thus, the enthusiasm of a non-Catholic audience for the Corpus Christi procession was as important in symbolic terms as the event itself. For the *l'Écho de l'Orient* reporter, Greek cooperation was a sign of a heightened level of tolerance. It momentarily dispelled any ill-feelings felt by other groups, and contributed to Catholic pride.

The public emergence of the Corpus Christi procession in 1842 also had implications beyond its initial performance because it established a precedent for Catholics to perform the event on a yearly basis. In the following decades, the French periodical press continued to report the celebration as a moment of tolerance and Catholic pride.[27] Through the remaking of Frank Street by demarcating territory, linking particular institutions, and positioning participants and observers, it periodically inscribed Frank Street as belonging to Catholics, at the same time demonstrating a harmony between the celebrating community and other communities. In addition, the Corpus Christi performance opened a political space for other groups to carry communal events into the public realm. Similar spatial keys and cues were used during the Greek Orthodox parade of St. George, which started at the church of that saint's name, also marched through a portion of Frank Street, and allowed a similar refashioning of space. On this day streets were decorated with the blue and white Greek flag. As church bells rang, devotees watching the event from their windows and balconies strewed rose water and flowers on the participants. The procession was led by two guards (*cavass*) of the Greek Club and the Greek consulate, wearing the national Greek costume. They were followed first by students, professors, male and female teachers, and the dean of Greek schools, marching in rows of four, and then by the Greek Orthodox clergy in gold and red and gold and blue vestments. The priests held silver censers, a holy flag decorated with angels, and the icon of St. George in a gold-plated box. The procession moved to the sound of prayers, chants, and hymns, mixed with cries of "long live the Metropolitan!"[28] During the same St. George's day march the head of the procession turned to the participants at every stop, took off his hat, and acclaimed "Long live the grand sultan Abdulhamit!" followed by "Long live our grand King George I!" The crowd reciprocated with similar cries.[29] Such enactments were relatively common in the midcentury context. The establishment of an independent Greek Kingdom in 1821 was pivotal in nurturing feelings of

Greek nationalism among Ottoman Greeks. Even though the independent Greek Kingdom did not readily preempt or supplant the older allegiance of Orthodox Greeks to their *millet*, it provided an alternative political loyalty which coexisted with other allegiances. Frequent commemorative events confirm this fact. On the name day of King George of Greece, *Amalthea*, the local Greek newspaper in Izmir, reported that the event was celebrated "openly and freely with the Greek and Russian consuls and their full staffs in attendance. . . . [T]he names of the Greek king, Russian emperor and sultan Abdulaziz were memorialized several times in the service."[30] The display of parallel allegiances was a way of accommodating the conflicting political loyalties which stirred the everyday life of Ottoman Greeks. These acts allowed local Greeks to advance new national loyalties without directly challenging the authority of the Ottoman rulers.

Appropriating established religious occasions for nationalistic purposes was another way of making new loyalties visible. In the later decades of the nineteenth century, Greek Orthodox Easter rituals provided an important channel for the public display of emerging allegiances. Many visitors to Izmir described the patriotic dimensions of Easter celebrations, but saw them under different lights. The Izmir reporter for the *Levant Herald*, a major Istanbul paper, attended a midnight Easter ceremony celebrated by 10,000 to 12,000 people holding lit candles in the courtyard of St. Photini. The belfry was illuminated top to bottom, and after the metropolitan made his lay announcement, fireworks were set off and firearms discharged. Looking at the event through a secular lens, the reporter found the mixing of religious and nationalistic symbols to be unacceptable, and even offensive. "The chants were not appropriate to the solemn occasion, with national marches and nondescript selections out of place in a Christian church."[31] Walking through Izmir on an Easter day, Gaston Deschamps saw that the windows of Greek houses filled with curious people and "charming brunettes," who in honor of Greece had stuck blue flowers in their jet-black hair. The same Easter day he witnessed a half-religious, half-patriotic celebration with priests and *palikares* (Greek gunmen) marching through the streets under the protection of Turkish police, who conscientiously kept onlookers away as the procession approached.[32] In his picturesque depiction of the event, Deschamps glossed over its political meaning by presenting it as no more than an amusing scene consisting of the incompatible juxtaposition of priests, ruffians and police. Commenting on a similar Easter event, Louis de Launay remarked on the apathy of eight Turkish policemen, led by a sergeant, who listened calmly to the *zitos* (Greek acclaims) at the doorstep of St. Photini.

> All fireworks were strictly forbidden yesterday by a formal edict of Ottoman authorities. The Greek anthem was too: it goes without saying . . . but one should let young people the joy of innocently violating the law: this would sometimes prevent greater upheaval.[33]

Yet neither were the violations innocent, nor were the Turks apathetic, as Launay put it. Public display of symbols such as raising the national flag and chanting the national anthem were the tangible manifestations of nationalizing Ottoman Greeks. In the last decade of the nineteenth century, as state control increased under the Hamidian regime, Greek national symbols became a source of tension for Ottoman authorities. In an official note sent to the Sublime Porte, local authorities in Izmir inquired whether during Easter, Greeks should be allowed to raise their flag next to the Ottoman flag on top of the prominent belfry of the Cathedral of St. Photini.[34] The propriety of this symbolic act was an important issue of contention. Although Ottoman authorities were increasingly sensitive to such displays, they seem to have allowed festive events as necessary outlets for communal tension. Public celebrations were ephemeral moments. Their brevity facilitated their official sanction, but did not diminish their significance in mediating emerging political aspirations and relations of power. These performances established temporal territorialities over more widely shared public spaces. During such events Frank Street was transformed into a symbolically charged stage, and provided a critical site for local communities and Ottoman authorities to reconfigure their relations with one another in the changing political climate of Tanzimat Izmir.

IMPERIAL CELEBRATIONS AS PERFORMANCES OF UNITY

Ceremonies such as the Corpus Christi and St. George's processions, or the Orthodox Easter rituals, coexisted with celebrations symbolizing Ottoman state power, such as the sultan's birthday and accession to the throne. In 1844, a week after the Corpus Christi procession, *l'Écho de l'Orient* reported with great enthusiasm on the sultan's birthday festivities: "All the mosques, Greek and Armenian churches and synagogues resounded with hymns addressed to the Almighty for Sultan's long life."[35] To observe this occasion, the delegations of every community of Izmir gathered ceremoniously at their respective temples and celebrated the sultan's birthday according to their own rituals. At the Armenian Church, following the service, Bishop Mateos made a short speech, acclaimed by "long live the Sultan." At the same time, the Jewish quarter displayed a festive atmosphere, and after attending synagogue, Jews marched through the bazaar to the sound of Amen chants performed by students in their holiday attire.

Such a formal performance of unity may seem incompatible with events such as the Corpus Christi procession or the St. George's day parade. Although the latter centered around the interest of a single community and was about articulating its particular position in the social context of the city and the empire, the former crossed religious lines to promote a supranational Ottoman community. Yet such differing public ceremonies coexisted as an embodiment of the key tension inherent with Tanzimat reforms — the effort to create a unified political allegiance out of the diversity of the empire. Aiming to create unity in the face of the rising nationalistic demands of local groups, the central authorities gave new impetus to celebrations symbolizing state power.

In Izmir the sultan's coronation was regularly reported as a very ecumenical event. The decorations, illuminations, and forms of celebration all contributed to this sense of all-inclusiveness. A repertory of conventions and a range of visual and auditory techniques were established for symbolic action in public space. A few days in advance, the governor general would make an announcement to the consular body and the chiefs of the Armenian, Greek and Jewish communities about the upcoming imperial ceremony. Most preparations were then made the night before. Official buildings and residences of notables and merchants of all communities would be decorated and lit. And from early evening until midnight warships in the harbor would fire salutes at short intervals. Similarly, at dawn a double salute opening the ceremony would be fired from the battery, which would be reciprocated from the citadel. Given that the harbor and the citadel were at opposite sides of the city and that such salvos would be repeated throughout the day, such gunfire ensured the whole city would be made equally a part of the event.

Until the late 1860s, Izmir had had no formally designed space for such official ceremonies, and the physical layout of the city had not recognized the need for performances symbolizing state power. In the early days of the Tanzimat, even the space around the governor's palace had been unorganized, located at the end of the bazaar and flanked to the south by a U-shaped barracks built in the 1820s for military maneuvers. But in 1860, on the visit of Sultan Abdulaziz, a new governor's palace, the Konak, was erected in place of the old, dilapidated mansion that had served this purpose. The construction of the new governor's palace turned the area into a more prestigious space. On the occasion of the sultan's celebration in 1866, for example, it was reported that the residence of the governor "glittered like a mass of fire."[36]

On regular business days, the governor's palace was a center for administrative transactions. The streets leading to it and the esplanade in front of it were filled with scribes and petitioners. But on official days the area became a focus of festivities. The remodeling of this area was completed when the new quay was built, allowing the esplanade in front of the palace to extend north the whole length of the city and forming the first cordon. In the following years many consulates moved to the new quay, and on festive days the first cordon was turned into an official thoroughfare. Framed on one side by the consulates raising their flags and on the other by steamers of different nations anchored in the harbor and flying gay bunting, it became a festive showcase.

Unlike other public ceremonies, which were played out in more bounded areas of the city and targeted to specific institutions central to the interests of single groups, the decorations for sultan's celebration extended throughout the city, overriding everyday divisions and boundaries. Every year *Hizmet*, the local Turkish newspaper, dedicated a special issue to the event, enumerating the buildings, shops and houses that had been illuminated. The governor's palace, barracks, battery, customshouse, guardrooms, Greek and Armenian bishopries, Ottoman Imperial Bank, and other prestigious buildings dressed their façades with garlands, while the consulates of France, Italy, Russia, Austria, Belgium, and other foreign countries displayed their flags.[37] Meanwhile, vessels in the port were decorated with banners during the day and with candles at night, and major thoroughfares such as Frank Street, Kemeraltı Street (the main artery within the bazaar), and other large streets had their shops and houses decorated with myrtle and evergreens. In many places the sultan's portrait, wreathed in flowers, occupied the place of honor in the decorations, and thousands of many-colored lanterns were lit when the night set in. The list of decorated buildings included hotels, theaters, and coffee shops on the quay (first cordon); stores, taverns and the Armenian bishop's house on the Street of the Great Taverns; the chief rabbi's house on Street of the Synagogues in the Jewish quarter; and the houses of several Muslim notables. The remaking of the city spaces through decorations, illuminations, fireworks and gunfire at such occasions imbued the spaces of the city with a new meaning, reinforcing feelings of a unified populous.

Members of various religious and national communities not only experienced, but also performed this coexistence by actively participating in the event. In the morning consuls in full uniform, accompanied by their *dragomans*, called on the governor general to present their congratulations. They were followed by civil and military authorities, heads of the religious communities (Muslim dignitaries, the Greek metropolitan, the Armenian bishop, the chief rabbi), and chiefs of the public administrations. In 1864 it was reported that the governor, in costume and wearing his *Osmaniye* decoration, made "a welcome in the most distinguished manner and responded with heartfelt words to the speeches delivered by the Greek Archbishop and the Armenian Bishop for the occasion."[38] Students of Ottoman, Greek, Armenian and Jewish schools came to the main hall of the governor's palace and delivered speeches in praise of the sultan. These codified sets of actions, periodically performed, attempted to entrench feelings of unity and harmony in Ottoman domains.

In the evening hours, the animation increased, and the town assumed a festive appearance. Crowds thronged the quay to see buildings and vessels brightly illuminated with either gas lamps or colored lanterns. Participants of all classes and faiths jammed the streets and public spaces where, ordered by the government, musicians played a variety of songs. Fireworks were set off continuously from the palace and the barracks. All the ships in the harbor were stationed at the esplanade in front of the palace and filled with curious eyes. "Muslims, Greeks, Armenians, Jews and foreigners of all nations would keep pleasant memories,"[39] concluded the reporter of *La Turquie*, who described how the ceremonies were performed in Izmir in 1866. In their visual and auditory remaking and inclusive participation, imperial celebrations staged a harmonious coexistence and inscribed it on the spaces of the city. In this way these events not only were the tangible expressions of the Ottoman political community advanced by Tanzimat reformers, but also helped formulate the ideology of a unified community.

RITUAL CALENDAR AND CYCLICAL "OTHERING"

Formal stagings such as the Corpus Christi procession and imperial events pointed to competing ways of performing community that emerged in the Tanzimat context. These events worked within a milieu characterized by a complex network of political allegiances, and their spatial enactment indicated an attempt to mediate multiple loyalties. By maintaining fluid and ambiguous boundaries between the celebrating community and other groups, or by temporarily denying social divides and instilling feelings of an agreeable coexistence, these performances sought to diffuse potential social tension.

The intercommunal dynamic, of course, had been complicated since before the Tanzimat era, and it was particularly charged at periods of religious observances. Muslim, Christian and Jewish ritual calendars were simultaneously observed. In addition, Orthodox and Catholic calendars were not aligned, and contained a great number of saints' days, many of which were kept as holidays. Based on the memoirs of the English consul, Paul Rycaut, who spent eleven years in Izmir in the mid-seventeenth century, Sonia Anderson wrote:

> Divines of every persuasion exchanged their views with an open-mindedness verging to heresy. The Greek bishop was in disgrace for neo-Jewish observances; the Armenian patriarch on a visit from Erivan in 1672 scandalized his congregation by wearing a pall and mitre presented to him by Pope Clement IX; the Jews produced a pseudo-Messiah who later turned Turk. Less extreme forms of syncretism were taken for granted. . . . Once a year the French consul and nation, accompanied by several Turks and one or two Jews, marched in procession to the Greek cathedral of St. Photin to celebrate Maundy Thursday by Orthodox rites.[40]

Seen through the eyes of an Englishman unaccustomed to such a level of cultural plurality, these intertwined practices indicate, nonetheless, the coexistence of a myriad of religious observances. In mid-nineteenth-century Izmir, most of these rituals were still kept, and often the calendar cycle of Jewish, Orthodox, Muslim and Catholic feasts coincided: either Ramadan coincided with Lent, or the Bairam was on Christmas or Easter. When the month of Ramadan coincided with Easter, as it did in 1860, the entire month could be dedicated to celebrations. Thus, on March 14 *La Turquie* reported:

> Business is slow. Jews are preparing for the Passover starting next Saturday. Catholics already in their Holy week are preparing for Easter. Greeks and Armenians are soon ending Lent, and their Easter would be a week before the Bairam of the Turks.[41]

The ritual calendar not only affected the pace of economic life in Izmir, but also corresponded to a relative positioning of individuals and groups with respect to each other. Regardless of the fact that different groups in the city observed different holidays, many rituals were made known to all by their daily implications. Bakeries and groceries owned by local Greeks or Armenians were closed on Sundays or major Christian holidays, forcing Muslims and Jews to make their provisions accordingly. Likewise, during the month of Ramadan, official transactions and municipal functions were suspended, because government agencies were closed.

Festive seasons not only were times of peaceful enjoyment, but also often brought periods of great religious excitement. For example, around Passover rumors often circulated that a Christian child had been kidnapped and slaughtered by Jews for religious purposes. Greek and Armenian rioters would then rush to the Jewish neighborhoods to harass passersby or break into stores. Although in every case the child would be found in a day or two, the same tragedy of "ritual murder" would be renewed the following year.[42] Events of the sort happened periodically between Passover and Easter, and it was often unsafe for a Jew to be seen near popular Greek quarters. For example, on March 26, 1873, the *Levant Herald* reported how three Jewish salesmen had been assaulted by a mob in a Greek neighborhood known as Tabakhane; how near the site of the Cafe Kivoto on the shore a Jewish child had been thrown into the sea and seriously wounded; and how close to the Caravan-Bridge another Jew had been beaten by a number of Greeks.[43] Potential religious tension periodically raised awareness about communal boundaries, forcing people to define temporary territorial limits. Between Passover and Easter, Jews viewed popular Greek neighborhoods as sites of danger, and avoided them. Such markings and withdrawals inscribed Greekness and Jewishness in particular spaces and neighborhoods, and cyclical acts of "othering" strengthened ill-feelings among local groups, marking space in a more violent and disharmonious way.

Greek Easter not only was a source of anxiety for the Jews, but it also affected other sections of the population as well. For example, the Easter festivities were a challenge to the European local elite's sense of public order. Easter was celebrated in a loud fashion in the Greek community, with pistol shots in the streets that often led to accidents. In 1876, there were 37 injuries and four deaths reported in popular Greek neighborhoods, marking these places as sites of peril to avoid temporarily.[44] Ottoman authorities always remained alert during the Easter season, and they increased their police force to monitor and maintain public order on the streets. In many ways seasons of ritual observance were moments of heightened awareness of the "Other" and reminders of differences between "us" and "them." Like formal religious parades, seasons of festival carried difference into urban space. Yet they inscribed space in inherently different ways. Unlike official events that carefully orchestrated communal differences and alliances between celebrating groups and other groups, seasons of religious excitement and spontaneous agitation marked everyday spaces in an abrupt manner. They periodically brought ephemeral, yet hostile, social divisions to the surface, suggesting the limits to which communal differences could be mediated through deliberate attempts and formal stagings.

MEDIATING DIFFERENCE AND COEXISTENCE THROUGH SPATIAL PRACTICES

In Tanzimat Izmir parades and ceremonial events worked within a complex social and political landscape. They juxtaposed, within the same spaces, conflicting social and political ambitions — both the efforts of the Ottoman authorities to promote a unified political identity, and the desires of local groups to advance their religious differences and express growing political allegiances to outside nation-states. Although this conflict was in part inherent with the Tanzimat reforms, it was amplified by the long-standing religious heterogeneity of Izmir. Through their spatial and dramaturgical strategies, the performances discussed here opened a site to mediate and confront these multiple social and political tensions.

These spatial performances also pointed to the different ways in which issues of identity and difference were fought out in urban spaces. Moments of staging and *mise-en-scène* offered one venue to enact religious difference and national identity within an otherwise charged social environment. During parades such as the Corpus Christi and the St. George's processions, or the Easter ceremonies, celebrating groups appropriated particular spaces and inscribed them as their own. By refashioning everyday spaces, organizing participants, and targeting particular audiences, such groups not only performed their belonging to the religious community, but also advanced emerging political aspirations. Counteracting these performances of difference were state-sponsored events aimed at constructing a unified populous out of socially diverse communities. On such imperial occasions, Ottoman authorities also employed spatial strategies to reclaim the space of the larger city and mark it with an image of sameness and unity. This was a way of momentarily dissolving traditional inequalities between subject people. It was also a way to present a picture of social and political harmony.

The orchestration of public events was not the only way of playing out identity and difference in urban spaces, however. Periods of ritual observance provided an opportunity to perform religious difference — yet in a more spontaneous and violent way. Ritual seasons and times of religious excitement cyclically revived long-standing prejudices and anxieties among groups, and constituted a continuing potential for social tension. These periods corresponded to moments of enhanced awareness of religious difference and to an abrupt reconfiguration of social and spatial boundaries within the city.

In the face of sociopolitical changes and complex intercommunal relations, the particular event with which I began this chapter, the Corpus Christi procession, may be seen as an embodiment of how local groups performed a sense of community, drew temporary territorial boundaries, and forged alliances to claim a place of their own in the plural social landscape of Izmir. Examining the social and the physical context of such public events provides an important tool for understanding the terms of cultural plurality and coexistence that gave shape and meaning to mid-nineteenth-century Izmir. It extends and complicates the history of non-Muslim Ottoman communities, which is too often flattened by means of generic story lines that emphasize structural and top-down transformations. At the same time, such an inquiry into the spatiality of the performances allows a more nuanced understanding of urban spaces. The spatial strategies deployed through these celebrations opened up, and sometimes closed down, different ways of experiencing everyday urban spaces and provides a reminder that urban spaces are not uniform stages, but multilayered sites. Such spaces acknowledge different actors, aspirations, and relations of power and allow for different trajectories to coexist simultaneously.

NOTES

1. *L'Écho de l'Orient*, May 27, 1842.

2. M. Rubin, *Corpus Christi: The Eucharist in Late Medieval Culture*, Cambridge, Cambridge University Press, 1991, p.243.

3. *L'Écho de l'Orient*, May 27, 1842.

4. Officially, non-Muslim Ottoman subjects were divided into four and later five nations as *millet-i khamsa* (the five communities), governed under the control of the Sublime Porte by their by respective Patriarchs and the Grand Rabbi of the Israelites. These nations consisted of Orthodox Greeks, Armenians (who split in 1829 as Gregorians and United Catholic Armenians), Latins, and Jews. These communities were not homogeneous in themselves. They revealed ethnic and linguistic differences. The Orthodox Greek *millet* consisted, for example, of Serbs, Romanians, Bulgarians, Vlachs, Orthodox Albanians, and Arabs. See J.H.A. Ubicini, *Letters on Turkey: An Account of the Religious, Political, Social and Commercial Condition of the Ottoman Empire*, London, John Murray, vol.II, 1856.

5. Non-Muslims had to pay a special poll tax in return for the protection granted them by the sultan, their eligibility for high administrative posts was limited, and they were forbidden to carry arms. There were also social emblems, such as restrictions on the type of clothing they could wear and on the color and type of their residences, that marked them as different from other groups. See H. Gibb and H. Bowen, *Islamic Society and the West*, London, Oxford University Press, 1960, vol.II.

6. See S. Slyomovics, "New York City's Muslim World Day Parade," in P. van der Veer (ed.), *Nation and Migration: The Politics of Space in the South Asian Diaspora*, Philadelphia, University of Pennsylvania Press, 1995; M. Ryan, "The American Parade: Representations of the Nineteenth century Social Order," in L. Hunt (ed.), *The New Cultural History*, Berkeley, University of California Press, 1989; and S. Davis, *Parades and Power: Street Theatre in Nineteenth Century Philadelphia*, Philadelphia, Temple University Press, 1986.

7. D. Goffman, *Izmir and the Levantine World, 1550-1650*, Seattle, University of Washington Press, 1990, p.139.

8. Original text of the *Hatt-ı Humayun*, February 18, 1856, in G. Aristarchi Bey, *Recueil des lois, réglements, ordonances, traités, capitulations et autres documents officiels de l'Empire Ottoman*, Constantinople, Imprimerie Frères Nicolaïdes, 1874, vol.II, p.15.

9. Europeans and consular interference in local municipal affairs were continually discussed in the early 1860s. *Levant Herald* reported several debates on this issue.

10. *Levant Herald*, May 2, 1866.

11. Three governors were successively appointed in 1872: Mehmet Sadık Pasha, Mehmet Sabri Pasha, and Hüseyin Avni Pasha. Two were appointed in 1873: Mustafa Süreyya Pasha and Ahmet Hamdi Pasha. One was appointed in 1874: Ahmet Rasim Pasha. And four were appointed in 1875: Hüseyin Avni Pasha, Kayserili Ahmet Pasha, Ahmet Esat Pasha, and Hurşit Pasha.

12. On the civil organization of the Catholic (or Latin) nation, see Ubicini, *Letters on Turkey*, pp.375-392.

13. A. Schlicht, *Frankreich und die Syrischen Christen*, 1799-1861, Berlin, 1891, p.155. Also cited in G. Bozkurt, *Gayrımüslim Osmanlı Vatandaşlarının Hukuki Durumu*, Ankara, Türk Tarih Kurumu, 1989, p.187, n.68.

14. "Frank" is a generic name given by Turks to Europeans, regardless of their nationalities.

15. Business directories of 1890s and insurance maps of 1905 showed that owners of shops and residences in the Frank area belonged to different groups.

16. C. Reynaud, "D'Athènes à Baalbeck," in J.-C. Berchet (ed.), *Le Voyage en Orient: Anthologie des voyageurs français dans le Levant au XIXe siècle*, Paris, Robert Laffont, 1985, p.50. Date of travel to Izmir: 1844.

17. Sometimes also called *frenkhane*, literally "residence of the Franks."

18. K. Politis, *Yitik Kentin Kırk Yılı*, Istanbul, Marenostrum, p.36.

19. *Journal de Constantinople*, July 27, 1860; excerpt from *Impartial*, July 20, 1860.

20. Comte J. d'Estourdel, *Journal d'un voyage en Orient*, vol.1, Paris, 1844, p.196.

21. The French college run by the Lazarists was first established in 1802. It subsequently burned, and was rebuilt. In the 1880s, it was moved from Rose Street to Frank Street.

22. The brotherhoods active in Izmir were the Lazarists, Recolets, Capucins, Dominicans and Mechitarists.

23. *L'Écho de l'Orient*, May 27, 1842.

24. In the late 1890s, authorities in Istanbul asked for the removal of Muslim students from schools under French protection. Başbakanlık Ottoman Archives, Yıldız Esas Evrakı, Sadrazam Kamil Pasha, nos.86/13 1283 and 86/15 1427.

25. George Rolleston wrote that it was rare for a Catholic to intermarry with a Greek family, and that priests employed all means in their power to prevent such unions. Ill-feelings also existed between Armenian and Greek populations, but more because of commercial competition than religious antipathy. See Rolleston, *Report on Smyrna*, London, 1856, pp.28-48.

26. *Levant Herald*, September 27, 1869.

27. In the late 1860s, the itinerary would be more powerfully anchored, because the march would begin not at the College of the Propaganda, but next door, at the newly erected and sumptuous St. John's Cathedral. *La Turquie*, June 7, 1866.

28. Politis, *Yitik Kentin Kirk Yili*, p.72.

29. Ibid., p.72.

30. *Amalthea*, April 28, 1867; cited in G. Augustinos, *The Greeks of Asia Minor: Confession, Community, and Ethnicity in the Nineteenth Century*, Ohio, Kent State University Press, 1992, p.242, n.23.

31. *Levant Herald*, April 28, 1877.

32. G. Deschamps, *Sur les routes d'Asie*, Paris, Armand Colin, 1901, pp.152-53. Date of travel to Izmir: 1887.

33. L. de Launay, *Chez les Turcs de Turquie*, Paris, Edouard Cornely, p.33. Date of travel to Izmir: 1887.

34. Başbakanlık Ottoman Archives, Istanbul, Yıldız Esas Evrakı, Sadrazam Kamil Pasha, no.86/18 1791.

35. *L'Écho de l'Orient*, June 15, 1844.

36. *La Turquie*, July 12, 1866.

37. *Hizmet*, September 1, 1887.

38. *Journal de Constantinople*, February 4, 1864.

39. *La Turquie*, July 12, 1866.

40. S. Anderson, *An English Consul in Turkey: Paul Rycaut at Smyrna, 1667-1678*, Oxford, Oxford University Press, 1989, pp.8-9.

41. *Journal de Constantinople*, March 14, 1860.

42. A. Galante, *Histoires des Juifs d'Anatolie*, vol.2, Istanbul, M. Babok, 1939.

43. *Levant Herald*, March 26, 1873.

44. The newspaper article referred to the neighborhoods of St. Photini, St. Jean Cathedral, St. Dimitri, St. Catherine, and St. Mary, as well as Bahçeli and Scalabrini Streets and the Street of the Great Taverns. *La Turquie*, April 26, 1876.

3

California Chinatowns: Built Environments Expressing the Hybridized Culture of Chinese Americans

Christopher L. Yip

The California Chinatowns born in the late nineteenth and early twentieth centuries expressed the need of Chinese immigrants to create built environments that could serve their culture and society. For a number of reasons, including the radically different nature of North American urban environments from those in China and the need for the immigrants to fit into a relatively hostile society at large, this process did not involve a simple transplantation of traditional Chinese buildings and vernacular settings. Instead, the Chinatowns of California embodied communities in which common North American building types and environments were transformed to meet a new set of needs **(Figure 3.1)**.

The social position of Chinese immigrants in the new land was an important factor that led to the creation of such hybrid physical environments. A process of adaptation was forced upon them by their status as cheap, imported labor at the bottom of the social order, for whom the boundaries to upward mobility were set both by racial discrimination and legal sanction. But the development of California Chinatowns must also be understood in terms of global factors. Chinese migration to the United States was part of a general pattern of migration set off by Western expansion and the impact of nineteenth century Western imperialism. Chinese migration to North America also occurred at a time of crisis and internal decline in China. During the nineteenth century, the pressures generated by China's explosive population growth and stagnant economy were only exacerbated by a conservative political establishment unable to fully grasp the critical nature of the situation. And in the Pearl River Delta in China's south, where the population explosion of the eighteenth and nineteenth centuries was particularly acute, the effects of declining per-capita agricultural acreage and increasing farm tenancy were worsened by midcentury droughts.[1] The poverty and social unrest that resulted from these forces acted as a strong factor pushing people to seek their fortunes abroad.

Figure 3.1. Grant Avenue, San Francisco, California. (Photo by author.)

At the same moment in history, the Western powers were busy incorporating areas of Asia into their economic spheres, and in many cases placing once-sovereign peoples under direct political control. In China, the British used opium that was produced in its Indian colonies as a primary trading commodity. This led to increased internal problems (based on massive drug addiction), and created a dramatic shift in the balance of trade in favor of Great Britain and the other countries (such as the United States) that benefited from the opium trade. Eventually, friction between China and these Western nations exploded into the first Opium War (1839–1842), in which Great Britain decisively defeated China in the name of "free trade." Thus began the process by which Western nations dismantled Chinese sovereignty over important port cities and economic resources. These incursions by Western powers only worsened China's political instability and social turmoil, especially in coastal southern China, where the vast majority of immigrants to the United States originated.

For centuries the peoples in the southern coastal provinces of China had been engaged in trade and migration to and from Southeast Asia (known in

China as Nanyang).[2] Connections between the overseas Chinese communities and South China were important to the maintenance of extended family ties that stretched across the South China Sea from the trading ports of Nanyang to ancestral homeplaces. They were also important to the promotion of regional commercial networks.

In Southeast Asia the presence of Chinese merchants had several benefits for the colonial powers. First, whereas the local population was largely rooted to the land, with a high proportion engaged in agriculture, the Chinese were not landed members of the societies to which they immigrated. Instead, they turned to trade and commerce, and could serve as middlemen between the local population and the Europeans who ran the government and the major enterprises of the colony (this was generally the case for the immigrant Chinese in the Spanish Philippines, French Vietnam, British Malaya, and Dutch Indonesia). Secondly, the Chinese served an important role in the development of trade between colonies. The official trade of each colony took place between the colony and its metropole in Europe (or later, in the case of the Philippines, the United States); in fact, the transfer of wealth and commerce from the colonial hinterland to the metropole was the primary concern of colonial governments and the European businesses they privileged. Although some of the indigenous people within each colony learned the European language of the ruling power, this did not equip them to deal with other Southeast Asian colonies. And because the colonial powers were not interested in fostering trade with rival empires, it fell to the Chinese to conduct intercolonial economic activity. The Chinese were ideally suited for this, sharing both a written language and business culture with Chinese residing in other colonies.[3]

Ultimately, the Chinese experience with migration to Nanyang became a model for migration to California. The back-and-forth pattern of migration that developed with Nanyang was repeated in California. So was the establishment of chain migration, as families, villages and districts sent new immigrants to locations where known people had already gone. But where the Southeast Asian and California experiences differed radically was in the way the Chinese fit into the local economy and culture of the societies they entered.

Unlike their experience when arriving in the colonial-dominated countries of Southeast Asia, early Chinese immigrants to the United States found themselves in a sovereign country dominated by a Euro-American population which condoned the enslavement of peoples of African descent. The United States was also in the last stages of a genocidal war to disenfranchise the Native American population. Chinese arriving in America were quickly relegated to low-paying, low-status jobs, both through general practices of racial discrimination and laws specifically enacted against them. This situation made it impossible for the Chinese population of America to assume the same prominent economic position they had come to occupy in Southeast Asia. Instead, the Chinese in the United States (and, of course, this included California) found themselves locked into a limited range of low-wage occupations. The only alternative was to enter businesses that mainly served the Chinese-American population, and in this sphere there were limited opportunities.

THE CALIFORNIA CONTEXT

The 1849 Gold Rush in California diverted a portion of the South China migration from Nanyang to California.[4] Possibly as many as 95 percent of the Chinese immigrants during the first decade following the discovery of gold came from a few districts of Guangdong Province.[5] Early Gold Rush immigration established the basis for the later pattern of chain migration that would bring the Chinese to California and other parts of the Pacific slope of North America. According to this pattern, later immigrants would set off for particular locations in California where they knew they could join relatives or acquaintances from their native villages.

San Francisco's Chinatown was the first stop for many of these immigrants, and it remained the largest Chinatown in the United States until the late twentieth century. In a way, the San Francisco Chinatown served as the metropole for Chinese in California, and it was the first place in which Chinese conceptions of place-making ran into North American patterns of urbanization. The hybridized strategies of adaptation that developed in San Francisco tended to diffuse out to the other Chinese communities across California and other areas of the Pacific slope. This included the spread both of social institutions and strategies for adapting North American building types to Chinese-American uses.

The need to adapt to the California context and develop strategies for the shaping of places was made necessary by the weak social, economic and political status of immigrant Chinese. As a highly discriminated-against minority group, their chances of imposing their architecture and concepts of place-making on the new environment was out of the question. Hybrid solutions that adapted to the American context posed the only reasonable approach.

Although Chinese-American settlements in California made use of common American building types that existed at the time, the conditions of daily life generated a separate settlement reality in Chinese communities than elsewhere. The special physical aspects of Chinese-American communities were shaped by several factors: first, the immigrants' cultural framework; secondly, the demographic structure of the Chinese-American community; and thirdly, discrimination and legal barriers raised against the Chinese in the United States.

Because immigration to California was conceived of as a temporary sojourn abroad to acquire enough wealth to return to China and establish financial security for oneself and one's family, the early waves of immigrants maintained strong ties to their native places. The young men who risked the arduous journey and unknown conditions in a foreign land were often married men whose wives and children had remained in their home villages. It was a common practice for parents to arrange a marriage for their son before allowing him to go off to "Gold Mountain." It was hoped the newlyweds would quickly produce a son to continue the family line, and while the son was gone, the daughter-in-law would add to the labor pool of the family. The immigrant's loyalty to his family and homeplace would also be greater if he knew he was leaving behind a wife and child, thus making it more likely he would send regular remittances of money and ultimately return to the homeplace. This process of bonding to a family in China before setting off for America helped make North America seem like an unlikely permanent home for young male immigrants. Ironically, the majority of Euro-American Gold Rush pioneers had a similar view of their migrations to California. In fact, Italian immigrants to San Francisco continued to view the city as only a temporary home for many years after their arrival.[6]

The Chinese immigrant population was overwhelmingly male, which was not unusual for a pioneer population. Women and children were rarities in the first decades of immigration, and remained a small portion of the Chinese-American population throughout the nineteenth century. The 1860 U.S. Census showed a ratio of about nineteen Chinese males to each Chinese female.[7] This represented the extreme outcome of a common migrational phenomena, by which men leave their families in an ancestral homeplace to seek wealth overseas, with the intention of returning once they have achieved financial success.

The demographics of the Chinese population of Gold Rush San Francisco — and by implication other parts of California during the early 1850s for which statistical information is not available — was not uncharacteristic of the city's population as a whole. During the first half of 1849, about 10,000 people landed in San Francisco, including only about 200 women. In 1850, about 36,000 people landed, including only some 2,000 women, a high percentage of whom were prostitutes.[8] Among Italian immigrants to San Francisco from the 1860s until the 1880s, the vast majority were men. And as late as 1892, only 17 percent of the Italian population of San Francisco were women.[9] Yet, unlike most other immigrant groups, a significant imbalance between males and females continued in the Chinese-American population into the mid-twentieth century.[10]

As the location of such "bachelor" societies for many decades, the Chinatowns of California thus became structured around the work and social institutions that catered to a predominantly single male population, rather than serving as a place for the transplantation of the extended family and other structures of Chinese life. The male social order of Chinatown was composed largely of laborers who sought whatever employment they could find. But, unlike laborers of European ethnicity, they were barred from all unionized jobs and from most forms of heavy manufacturing, because California unions were in the forefront of the anti-Chinese movement. Large numbers of Chinese laborers, however, did find work in the most dangerous areas: railroad construction, mining, agricultural work, and levee construction; and they returned to Chinatown when their work ended and they needed to find another job. Other Chinese worked in small Chinatown factories.[11] Still others worked in the small businesses of Chinatown, such as dry-goods stores and restaurants. At the bottom of the Chinatown social hierarchy was the body of Chinese laborers. Just above were the labor bosses who contracted for jobs and brought teams of laborers to job sites. At the top of the Chinatown social order were the few business owners and the heads of the various Chinese associations (most of whom were businessmen).

If the male residents of California Chinatowns struggled under the burden of severe racial discrimination, the few Chinese female residents suffered the double burden of racial and sexual discrimination. Racism bound such women to Chinatown, and patriarchy held them in their dwellings or brothels subservient to men. The Confucian "Three Obediences" required a woman to obey her father in childhood, her husband upon marriage, and her eldest son if she became a widow. In China it was common for many women to be trapped in the home, or within a working zone in and around it. Travel by women abroad was rare, both because it was customary for women to stay at home when their husbands sought work abroad, and because Chinese law forbade female emigration until 1911. In this regard, Chinese law (though loosely enforced) was markedly different from laws that affected emigration from European countries. When a Chinese female made the journey to the United States, it was almost always as the property of a man. Thus, almost all women who immigrated to California were either prostitutes, indentured domestic servants (*mui tsai*), or wives joining their husbands. A large proportion sailed to San Francisco and remained there. In 1860, about 37 percent of all Chinese females in the United States lived in San Francisco, and by 1900, the percentage had risen to 47 percent.[12] If there were relatively few Chinese women in San Francisco, the imbalance between men and women was even more extreme in the other Chinatowns of California.

Among the Chinese women in California, the largest group were prostitutes. Often such women were held in bondage by contracts that required them to work for four to six years. Due to the dangers of this employment and abuse at the hands of their owners, few lived out the period of their contracts. Many had originally been kidnapped or tricked with false promises of marriage, only to find themselves in a Chinatown brothel.[13] According to Judy Yung's estimates, about 85–97 percent of Chinese women in San Francisco in 1860 worked as prostitutes. By 1870, the percentage had dropped to about 71–72 percent, and by 1880 it was about 21–50 percent.[14] These percentages provide a suggestive measure of the prevalence of prostitution among Chinese-American women in California during the late nineteenth century. In fact, prostitution as a major occupation only died out in the 1920s.

After prostitutes, *mui tsai*, domestic servant girls, made up the next, slightly higher level of Chinatown's female social hierarchy. In traditional China, young girls might be sold by poor parents to other families. These girls were expected to work as domestic servants for food, clothing and shelter, but no pay, until they reached a marriageable age, at which point they would be married off. Under this system, *mui tsai* could either be treated as a "little sister" and a member of the family, or worked mercilessly and occasionally resold to another family or into prostitution. The *mui tsai* in California served the wives of successful merchants and leaders of the Chinese community. The *mui tsai* system functioned in California until the 1920s.

Immigrant wives constituted the top level of Chinatown's female population. Their numbers were few in the early years, but steadily increased. Judy Yung suggests that between 1870 and 1900 the percentage of married women in the Chinese female population of San Francisco grew from only 8 to about 62 percent.[15] A small number were wives of merchants and community leaders, who could spend their lives in Chinatown apartments raising children and chatting with Chinese neighbors. Most, however, were married to modest business owners or laborers. They lived lives of constant toil, their time divided between raising a family in a tenement room or apartment, and work: sewing, washing clothes, clerking in a shop, or laboring in some small-scale processing industry. It was not uncommon for a working-class woman to find herself running a family of half a dozen in the confines of one room in a Chinatown tenement building. Her mobility was limited to the confines of Chinatown, outside of which lurked the fear of racial assault. Even the streets and alleys of Chinatown brought the fear of sexual assault.

Under such conditions of patriarchal domination within the Chinese-American community and discrimination from the society at large, women had little power to shape their place within the Chinatown landscape. Usually, decisions were made for them by the men who controlled their lives.

SYSTEMATIC DISCRIMINATION

Given such circumstances, some hybridization of the physical environment was unavoidable. Cultural differences alone between the Chinese and the Euro-American milieu would have required some adaptation of traditional forms. Thus, hybridization would have taken place even if the Chinese immigrants had been attempting only to maintain "Chinese" institutions within a relatively benign new social and institutional setting. But in California Chinatowns, hybrid physical environments were as much a response to severe discrimination as they were to cultural difference. Social prejudice, political disenfranchisement, and exclusion from further immigration formed the external forces that encouraged internal cohesion within the Chinese-American community. Chinese-American communities turned inward for social support and relief from the oppressive behavior of the majority society.[16]

Overt hostility to the presence of the Chinese began in the gold fields of California soon after their arrival, and it continued in the western United States in various forms well into the middle decades of the twentieth century. In the early years, the Chinese were routinely driven from the gold fields and physically abused. These patterns persisted and worsened in the 1880s, when Chinese Americans were attacked by anti-Chinese rioters in 34 different communities across California.[17]

National, state and local laws and ordinances limiting various aspects of Chinese American life were numerous. A few examples will suffice. Beginning in 1790, about sixty years before any significant Chinese immigration occurred, a federal law blocked Chinese from becoming naturalized citizens of the United States. The California statute of 1850 (the year 450 Chinese were reported to have entered through the Port of San Francisco) blocking African Americans and Native Americans from testifying against Euro-Americans in court was extended to the Chinese by the State Supreme Court in 1854 (at which time the annual Chinese immigration had reached many thousands). This ruling essentially made it all but impossible for a Chinese immigrant, confronting a Euro-American in either a civil or criminal proceeding, to receive justice, thereby stripping them of protection by the courts. In 1880, the State of California enacted a law prohibiting intermarriage between Chinese and Euro-Americans. These and many other laws, ordinances, and court decisions tended to remove the protection of the laws from the Chinese, while isolating and stigmatizing them as a group and opening them up to abuse and unfair treatment.

Laws were also employed to halt the growth of the Chinese-American population. Beginning in 1882, the federal Exclusion Acts made it increasingly difficult for Chinese to enter the United States.[18] The Act of 1924 went so far as to block the right of American citizens of Chinese ancestry to bring their foreign-born wives and children into the United States. The Exclusion Acts were passed during a period in which the Chinese-American population was still overwhelmingly male. In 1880, the male-to-female ratio was about twenty-one to one.[19] Beginning in 1882, the Exclusion Acts, emigration back to China and mortality combined to generate a decline in the Chinese-American population which continued until the 1920s. This froze the social structure of California Chinatowns in the first-generation patterns of pioneering male communities.

The various forms of discrimination (legal and otherwise) had the effect of restricting Chinese employment to a narrow range of occupations. Chinese were routinely driven from the gold fields during the Gold Rush, and the California Foreign Miner's taxes of 1852, 1853 and 1856 were directly aimed at removing the Chinese from gold mining. Furthermore, the California Constitution of 1879 prohibited Chinese employment by state, county or municipal governments; local ordinances pushed the Chinese out of the fishing industry; and discriminatory practices kept the Chinese out of the professions, white-collar employment, and many unionized forms of labor. As a result, Chinese workers concentrated in mining, railroad construction, agricultural labor, textile and clothing manufacture, restaurant work, household service, and laundries.[20] For the most part, the only Chinese who could be found dwelling outside of Chinatown were live-in household servants and laundry workers, who lived in their business establishments in Euro-American residential districts.[21]

For comparison, one might compare this situation with that of San Francisco's Italian population (who generally experienced less hostility than Italians who arrived in New York). Italians in San Francisco were able to move quickly out of their two original enclaves into every district of the city, and intermingle with other Euro-American groups. Italians were also higher in the social order than either Chinese or Mexican immigrants. Italians also benefited from a greater access to the economic opportunities posed by San Francisco's early stages of industrialization and urban development.[22]

THE LOCATION OF CHINATOWNS

The structure and location of Chinatowns were related first and foremost to the impacts of political, social and economic discrimination, and secondly to cultural factors. Segregation of the Chinese was the norm in most towns and cities in California. Chinese Americans were not allowed to purchase housing in Euro-American neighborhoods, and as a largely single male population, they had little need for houses built for nuclear families anyway. The few Chinese American families that did exist were barred from Euro-American residential neighborhoods, and were limited to low-rent districts in or near commercial and industrial areas.

As many geographers and sociologists have noted, the American city tends to have a core in which the central business district (CBD) is located. In its early years of development, this area often heterogeneously combined commercial, industrial and transportation facilities. But as the city grew, these elements tended to separate out into more homogeneous zones. Residential hotels and other forms of low-cost housing interspersed with other uses to form a ring around the CBD, offering inexpensive housing to workers who needed to be near sources of employment. Out beyond this ring was more established working-class housing, and further out yet lay the neighborhoods and suburbs of the middle and upper classes. Within this contextual order, the Chinese found themselves neither socially nor financially capable of locating in the suburbs or outer neighborhoods of the city (with the exception of laundries). Realtors would not sell to the Chinese, and Chinese isolated in neighborhoods of other ethnic groups often suffered from hostility and mistreatment. The California Alien Land Law of 1913 outrightly banned most Chinese Americans (specifically, those born outside of the United States) from owning land. The Chinese had no way of altering this situation, because first-generation Chinese Americans were barred from becoming naturalized citizens until after World War II, leaving the vast majority of adults in Chinatown without the vote or any other form of political power. Furthermore, during the nineteenth century the Chinese were mostly renters, because the discriminatory practices of the California Alien Land Laws blocked them from buying property. For instance, the whole Chinatown of the Sacramento Delta town of Locke, which was constructed in 1915, stood on land leased from a farmer, George Locke.[23]

Chinatowns tended to occupy the mixed-use, low-cost housing zone next to the CBD. There, the Chinese occupied a jumbled building stock, which had been modestly transformed to include the elements necessary to support an overwhelmingly male community. In the case of San Francisco, Chinatown developed along Clay Street and Grant Avenue (then known as Dupont Street) near Portsmouth Square. This had been the core of the original settlement of Yerba Buena, but as the city grew, tidal lands had been filled in, and the CBD had moved eastward to be near the wharves, factories and warehouses along the new shoreline. The older, mixed building stock that had survived the many fires, or that had been rebuilt after various fires, was no longer in a prime commercial location, and was not deemed suitable by most Euro-Americans, who could afford to live elsewhere.

Figure 3.2. China Alley, Hanford, California. (Photo by author.)

Similarly, the Chinatown in Sacramento lay near the once-busy riverfront. The small Chinatown in Hanford was located in an alley near the railroad tracks, away from the central park and government buildings **(Figure 3.2)**. San Luis Obispo's Chinatown grew up at the edge of the CBD, next to the old Spanish mission, and served Chinese railroad laborers. Often Chinatowns formed a "skid row" which was looked down upon by the rest of the city. As Paul Groth has noted, "Common characteristics of the people inhabiting these districts were unskilled and low-paid work, living outside of a private household of any sort, and being ostracized from the rest of the city."[24] This described the situation of the Chinatowns of California.

It was not unusual for city fathers to move a Chinatown if and when they saw fit to do so. The Los Angeles Chinatown was torn down and moved a number blocks to make way for a new railroad station. Similarly, Chinatown in Oakland was moved four times.[25] The movement of Chinatowns was made easier by the low rate of Chinese-American property ownership.

COMMON CHINATOWN BUILDINGS AND USES

A rather limited range of architectural forms tended to dominate the typical California Chinatown, the most common being the multiuse commercial structure of the late nineteenth-century city. Within these buildings various uses would be housed. In small towns, a Chinatown would consist of one- and two-story wooden structures, sometimes including some single-use structures, such as in Weaverville, Courtland and Locke. In general, the range of building types and uses found in Chinatowns was related to the cultural patterns of the community and the conditions posed by the urban setting and the local building industry. Mixed-use structures housing some combination of commercial activities, residential hotels, association rooms, and temples were common. Often buildings were owned by an absentee landlord who rented space to the Chinese.

When compared with other ethnic groups in California, the Chinese placed a much lower emphasis on housing, a result of their economic situation and social status. Because the Chinese population was mainly composed of single men who often sent significant portions of their incomes back to families in China, there was relatively little interest within the Chinese community for spending money on apartments or houses — and, in fact, there was little money to waste. Even if a Chinese person wanted to own a house, there was little available space in most Chinatowns; and it would be unusual for a community to allow a Chinese to own land or lease and build outside a Chinatown. Only native Chinese Americans could legally buy property in any case, and then only if they could find a seller. By contrast, Chinatown property was viewed as a "cash cow" for Euro-Americans, demanding little or no maintenance.

The majority of the Chinese-American population found housing in residential hotels, or slept at their places of work. In the hotels, one room could hold multiple beds, and men might sleep in shifts to keep individual costs down. At work, beds could be placed in storage rooms. Under such conditions, contacts with kin (fathers, uncles, brothers and cousins); members of one's ancestral village, district, or dialect group; or friends were best conducted outside the dwelling (sleeping) space. Many Chinese also ate their meals in the numerous restaurants of Chinatown districts.

In Chinatown a residential hotel could be tucked away in almost any available space. Many residential hotels, composed of single rooms with toilet and washing facilities at the ends of hallways, could be found on the upper floors of multistoried brick commercial buildings. Sometimes rooms only had windows onto light wells, and it was common for more than one man to occupy a one-room unit of between 6 x 8 to about 10 x 15 feet in size. Larger hotels might have cooking facilities at the end of a hall or in one or two designated rooms.[26] These establishments often varied in occupancy based on seasonal fluctuations in work, as laborers working in agriculture or levee and railroad construction returned from rural work camps at the end of each season to wait for their next job. Families could be found in one or two adjoining rooms. As the number of children increased, the hallways became play areas, and the floors of large residential hotels took on the character of neighborhood streets. Because the owners of Chinatown properties knew that they served a captive market of mostly single men who were not able to find housing elsewhere, many residential hotels were either old and dilapidated or roughly finished.

Residential hotels were a common housing type across the United States from the late nineteenth through the mid-twentieth century, and as Paul Groth has noted, Chinatown residential hotels represented a distinct subset of this larger phenomena. For instance, San Francisco in 1880 had about 15,000 residential hotel rooms, housing about 6 percent of the population. By 1910, the number had grown to about 65,000 rooms, housing about 10 percent of the population, a ratio that remained roughly constant until 1980.[27] Although the range of accommodations ran from "palace hotels" for the wealthy, to "mid-priced hotels" for middle-class families, to "rooming houses" for workers, and "cheap lodging houses" for the people at the low end of the economic scale, the Chinese were limited to those residential hotels in their section of the city and these were overwhelmingly at the lower end of the scale.[28]

Within California Chinatowns, the most elaborate buildings tended to be association buildings. These represented the organizations that dominated the community. Association membership was commonly based on kinship ties, surnames, village or district affiliations or dialect group. These organizations offered their members social contacts, job opportunities, links back to the ancestral homeplace, and could even function as quasi-governmental organs.[29] An association rarely occupied a whole building, however, even if it owned it. Commonly, the association rooms occupied only the top floor, while the ground floor was rented out for commercial activity that needed a street frontage. If there was more space available on middle floors, one might find either a residential hotel or some other commercial activity there.[30]

Associations identified themselves on the outside of such buildings by concentrating their use of exterior ornament and decoration on the association level at the top of the front facade. Common decorative elements were used to accentuate the importance of the association in the community, and included the use of curving eaves with pseudo temple bells dangling from the corners; red Chinese columns decorated with coiled dragons holding up the eaves of a top-floor loggia; and bright green, yellow-gold, and red color schemes. Standard cast-iron balustrades could likewise be transformed by paint and mounted light bulbs. Compared with the top, the middle-floor facade tended to be bland and undecorated, while the ground level was given over to shopfronts, often with bold signage and large glass display windows. Signage in Chinese added to the character of the building facade.[31]

Architects and builders responsible for the design, construction and remodeling of such buildings were all Euro-Americans. They would build the multiuse commercial structures for a landowner, who would then lease the property to Chinese tenants who would be allowed to remodel it to fit their particular requirements within boundaries established by the owner. The architects and designers knew very little about traditional Chinese architecture, and drew upon

Figure 3.3. Weaverville Joss House, Weaverville, California. (Photo by author.)

sources such as the early editions of Sir Banister Fletcher's *A History of Architecture*. They then used standard North American building materials, including tiles, iron railings, light bulbs, glass, and pressed-metal ceilings and walls to create "Oriental" designs. These were generally applied as surface treatments to the standard wood or brick commercial building.

Many Chinese immigrants brought local syncretistic religious practices to California, which combined elements from ancestor worship, local cults, Chinese mythology, Taoism and Buddhism. Temples tended to be dedicated to a primary deity, around whom others would form an entourage. The resulting pantheon might include various gods, goddesses, bodhisattvas, immortals, supernatural beings, and mythical and historical figures. Some of the popular deities were the God of War (Kuan Kung), the Goddesses of Mercy (Kuan Yin), and the Empress of Heaven (Tien-hou).[32] Temples in rural towns, such as the Weaverville Joss House and the Taoist Temple in Hanford, were constructed in wood or brick and could be built as free-standing structures **(Figure 3.3)**. In such structures, the main hall tended to be symmetrically arranged around a central axis, culminating in the main shrine, with the primary deity at its center. One or more altar tables would then be set directly in front of the shrine, where incense and offerings could be left. Secondary shrines could be added along the sides or in other rooms. Such temples might often be found in multistoried commercial structures, where they might displace an association from the top floor.

Figure 3.4. Ah Louis Store, San Louis Obispo, California. (Photo by author.)

Other activities sought whatever space was left in the buildings of Chinatown. Small factories could be found in basements, storage rooms, and the middle floors of buildings. Restaurants normally sought ground-floor street frontages, although some could be found along alleys and on second floors, where the rent was less. Brothels, gambling establishments and opium dens were consigned to basements and back rooms, where they would be out of sight of the general public and the police.

In addition to such structural characteristics, the strong identity of many Chinatowns owed much to signage that included the extensive use of Chinese characters, the use of standard American building elements in uncommon ways to mimic Chinese architecture, and the setting up of displays of produce and other goods for sale along sidewalks in front of shops. Chinatown was also a place where a high concentration of non-Euro-Americans were crowded together, increasing their visibility. Finally, Chinatowns had the allure of a "skid row," a place of mystery (cultural difference) and deviousness (working-class poverty, taken as moral imperfection). These visual and psychological elements helped account for the strong identities of California's Chinatowns **(Figure 3.4)**.

A CONCLUDING THOUGHT

Two points should be kept in mind with regard to California Chinatowns and the larger topic of this book — the conditions that create hybridized urban environments. First, the Chinatowns of California (and of the United States in general) were neither composed of direct importations from Asia, nor of standard building types left unchanged. Rather, they represented the transformation of existing U.S. building types to the needs of the Chinese American immigrant group. As such, they arose out of the interface between a particular immigrant group and the setting in which they found themselves. Secondly, unlike most other immigrant groups that came to the United States at the time, Chinese Americans were not able to disperse into suburban and residential neighborhoods as nuclear families within the first half of the century, and so they lived in communities that were often 80 to 90 percent Chinese. It was the very persistence of these clearly defined Chinatowns for many decades that further helped explain their strong visual identity in the urban fabric.

NOTES

1. F. Wakeman, *Strangers at the Gate: Social Disorder in South China, 1839–1861*, Berkeley, University of California Press, 1966, p.127.

2. C.P. Fitzgerald, *The Southern Expansion of the Chinese People*, New York, Praeger, 1972; and L. Pan, *Sons of the Yellow Emperor*, Boston, Little, Brown, 1990.

3. L. Crissman, "The Segmentary Structures of Urban Overseas Chinese Communities," *Man*, vol.2, no.2, June 1967, pp.185–204.

4. California was known to the Chinese as "Gold Mountain," and later became known as "Old Gold Mountain" after the Australian Gold Rush.

5. T.W. Chinn et al. (eds.), *A History of the Chinese in California*, San Francisco, Chinese Historical Society of America, 1969, p.20.

6. See D. Cinel, *From Italy to San Francisco: The Immigrant Experience*, Stanford, Stanford University Press, 1982.

7. C.L. Yip, "The Impact of the Social Historical Context on Chinese-American Settlement," in J. Lim (ed.), *The Chinese-American Experience: Papers from the Second National Conference on Chinese-American Studies*, San Francisco, Chinese Historical Society of America and Chinese Culture Foundation of San Francisco, 1982, pp.139–140.

8. F. Soule, J.H. Gihon, and J. Nisbet, *The Annals of San Francisco*, New York, D. Appleton & Co., 1855, pp.243,300.

9. Cinel, *From Italy to San Francisco*, p.163.

10. S.M. Lyman, *The Asian in the West*, Reno and Las Vegas, Western Studies Center, 1970, pp.27–28,79.

11. P. Chiu, *Chinese Labor in California*, Madison, University of Wisconsin Press, 1967.

12. J. Yung, *Unbound Feet: A Social History of Chinese Women in San Francisco*, Berkeley, University of California Press, 1995, p.24.

13. L.C. Hirata, "Free, Indentured, Enslaved: Chinese Prostitutes in Nineteenth Century America," *Signs: Journal of Women in Culture and Society*, vol.5, no.1, Autumn 1979, pp.3–29.

14. Yung, *Unbound Feet*, p.29.

15. Ibid., p.41.

16. S. Chan, *Asian Americans: An Interpretive History*, Boston, Twayne Publishers, 1991, p.45.

17. S.H. Tsai, *The Chinese Experience in America*, Bloomington, Indiana University Press, 1986, pp.67–70.

18. See E.C. Sandmeyer, *The Anti-Chinese Movement in California*, Urbana, University of Illinois Press, 1973.

19. Lyman, *Asian in the West*, p.79; 1880 U.S. Census.

20. P. Chiu, *Chinese Labor in California, 1850–1880: An Economic Study*, Madison, The State Historical Society of Wisconsin, 1967.

21. See P.C.P. Siu, *The Chinese Laundryman: A Study of Social Isolation*, New York, New York University Press, 1987.

22. S.L. Baily, *Immigrants in the Lands of Promise: Italians in Buenos Aires and New York, 1870–1914*, Ithaca, Cornell University Press, 1999, pp.223–228.

23. C.L. Yip, "A Chinatown of Gold Mountain: The Chinese in Locke, California," in T. Carter (ed.), *Images of an American Land: Vernacular Architecture in the Western United States*, Albuquerque, University of New Mexico Press, 1997, p.161.

24. P. Groth, *Living Downtown: The History of Residential Hotels in the United States*, Berkeley, University of California Press, 1994, p.152.

25. W.T. Chow, "The Reemergence of an Inner City: The Pivot of Chinese Settlement in the East Bay Region of the San Francisco Bay Area," Ph.D. diss., University of California, Berkeley, 1974, pp.116–118.

26. J.K.C. Liu, *San Francisco Chinatown Residential Hotels*, San Francisco, Chinatown Neighborhood Improvement Resource Center, 1980; and C.Y. Yu, "From Tents to Federal Projects: Chinatown's Housing History," in J. Lim, *The Chinese-American Experience*, pp.130–139.

27. Groth, *Living Downtown*, p.19.

28. Ibid., pp.26-167.

29. H.M. Lai, "Historical Development of the Chinese Consolidated Benevolent Association/Huiguan System," in *Chinese America: History and Perspectives 1987*, San Francisco, Chinese Historical Society of America, 1987, pp.13–51.

30. C.L. Yip, "Association, Residence and Shop: An Appropriation of Commercial Blocks in North American Chinatowns," in E.C. Cromley and C.L. Hudgins (eds.), *Gender, Class, and Shelter: Perspectives in Vernacular Architecture, V*, Knoxville, University of Tennessee Press, 1995, pp.109–117.

31. C.L. Yip, "Association Building," in P. Oliver (ed.), *Encyclopedia of Vernacular Architecture of the World*, Cambridge, Cambridge University Press, 1997, vol.1, p.770.

32. K.W. Wells, *Chinese Temples in California*, Palo Alto, R & E Research Assoc., 1962; rpt., 1971.

4

A Colonial Portrait of Jerusalem: British Architecture in Mandate-Era Palestine

Ron Fuchs and Gilbert Herbert

Throughout the ages — in fact, for millennia — Palestine has been one of the busiest cultural crossroads of the globe. Geopolitics have repeatedly made this poor and narrow stretch of land, caught between the desert and the sea, an arena where world powers have clashed and widely diverse cultural systems have collided and interacted. Its history has always evolved under the shadow of global forces, its own traditions ever formed within a state of cultural friction and diversity. Yet, if this fate could spell cycles of destruction, it could also generate great creativity. And friction and interaction have sometimes produced a synthesis of unique potency.

Ever since King David made it his capital at the beginning of the first millennium B.C., the city of Jerusalem has been a principal focus of Palestine's complex history. Both a small town on the verge of the desert and an *umbilicus mundi*, it has been at once provincial and cosmopolitan. It has, therefore, provided a microcosm of global conflicts, a mosaic of peoples, religions and cultures, and a place of multiple and contested identities. Jerusalem is undeniably a paragon of what may be termed "hybrid urbanism."

Yet a new chapter in Jerusalem's long history of cultural encounters began during World War I when British forces brought 400 years of Ottoman rule in Palestine to an end. The Jerusalem that the British troops entered in 1917 was a predominantly Oriental, traditional city. It had been dominated by Islam for twelve centuries (discounting the Crusader interlude), and the Haram, with its mosques — the navel of the city — proudly declared this fact. Its sixteenth century walls enclosed a dense urban fabric penetrated by winding alleys and divided into quarters by the major streets and bazaars, and its houses, clustered around inner courts, featured massive stone vaults characteristic of the vernacular tradition of the Judaean hills and central Palestine. Yet as provincial as Jerusalem had become during the Ottoman period, it had retained its diversity and cosmopolitan aspect. As a holy city to three world religions, its name was cherished by millions around the world. And as a site of holy places, it drew pilgrims, and sometimes immigrants, from the four corners of the earth. By 1917, its population included, beside Muslims, a large Jewish community (around half the population) and Christian communities of every denomination and ethnic origin.

The British arrived in Jerusalem during a period of increasing cultural diversity, when its traditional Oriental character was being challenged. Since the mid-nineteenth century, Palestine, like many other Ottoman provinces, had undergone a slow process of modernization and was increasingly exposed to Western technology and culture. Economic expansion and Ottoman reform had also brought the emergence of a bourgeois class that had acquired familiarity with Western ways and ideas. Non-Muslim communities had also begun to thrive, as restrictions on them were progressively relaxed, and the impact of non-Muslims on city life was being felt as never before. Modernizing elements within the Jewish community, later joined by Zionist activists, contributed to a Jewish cultural revival, and the European consulates and missionary-philanthropic organizations that had begun operating in Jerusalem around midcentury served as vehicles for the introduction of Western influence. The call of the *muezzin* now mingled with the chiming of church bells. But behind the new European institutions lay the ambition of the European powers to promote their interests in the city. Thus, competition between the powers for a foothold in Jerusalem reflected competing imperial drives for power in prewar Europe, just as surely as it demonstrated the lasting potency of the historic and religious significance of the holy city.

By 1917, all these developments were in the process of progressively transforming the cityscape. The bare land outside the city walls was gradually being filled by new construction, and a modern Jerusalem was slowly emerging, with the Jaffa Road as its main commercial artery. Neighborhoods of well-to-do villas and mansions were developing north and west of the city center, their houses adopting a new bourgeois style and forsaking traditional ways in favor of modern manners and techniques. Since 1860, the overcrowding of the Jewish quarter had also resulted in the construction of many Jewish housing estates around the city. Finally, the whole skyline of Jerusalem was being transformed by the monumental edifices and large compounds of European religious and philanthropic institutions, which had mushroomed around the intramural city (and in a few cases inside it). All these forces were to accelerate during the Mandate period following the British arrival, and Jewish immigration and Zionist initiatives were to play an especially dominant role in the development of the modern city.[1]

With the British occupation, Palestine found itself incorporated into a new context: the world-embracing system of the British Empire. This meant a new cycle of intercultural encounters for Palestine, but it also brought into focus the problem of the country's own identity. As the Ottoman Arab provinces passed to British and French control, the centuries-old political and cultural continuity of the Ottoman Levant was broken up into distinct entities, run as modern states. Thus, for the first time in many centuries Palestine re-emerged as a political entity with clearly defined international borders, and with Jerusalem as its capital.

But the task of governing the country demanded that the British Mandate administration not only establish civil institutions, but also define and construct a "national" identity. Even simple actions, such as the minting of coinage and printing of stamps, inevitably involved selecting or devising new symbols. Thus, the multilayered heritage of the land, its conflicting meanings and contested past, had to be sorted out and represented in a formula that fit its new status.

Building and architectural design was one sphere of British activity that particularly invited British statements of "Palestinianity." Indeed, one characteristic common to the work of all British architects who designed in Palestine during the Mandate period was a deep concern with the spirit of the place.[2] Even buildings representing British authority — in fact, especially those buildings — were conceived as architectural portraits of Palestine rather than obvious symbols of British dominance. British architecture of the Mandate period spoke of Palestine, not of Britain, and seemed to celebrate the indigenous, the traditional and the Oriental. This chapter explores the hybrid meanings of this British rendering of a Palestinian architecture.

COLONIAL REGIONALISM

British celebration of the landscape of traditional Palestine cannot be understood only in the specific context of Palestine. Unique as Mandatory Palestine may have been, it was still part of the system of the British Empire. Although not defined as a colony, Palestine was administered by the Colonial Office and by a bureaucracy trained in colonial methods, and a regionalistic preoccupation with the indigenous was a familiar phenomenon in the colonial architecture of the time. The Indo-Saracenic style of the British Raj in India and the sophisticated blend of Orientalism and Classicism evident in the government buildings of New Delhi are perhaps the best-known and best-studied examples. But the French *arabisance* in North Africa and Italian architecture in the Dodecanese were other notable examples. The meaning of this colonial Orientalism, or better, "colonial regionalism," has been discussed extensively in recent studies.[3]

The rationale behind such regionalism in colonial architecture can perhaps ultimately be traced to a paternalistic conception of the role of the colonial power, a conception holding that "all political power which is set over men ought to be in some way or another exercised ultimately for their benefit. . . . Every species of political dominion and every description of commercial privilege . . . are in a strictest sense a trust."[4] This theory of colonial government, the "doctrine of trusteeship" as it is sometimes called, regarded colonial dominion as a trust that entailed responsibilities on the part of the colonial power for the well-being of the ward nation. The adoption of indigenous motives in colonial representational buildings thus aimed at presenting the structures, and the institutions they housed, as conceived ultimately for the benefit of the colonial subjects. "We are trustees for India's intellectual and material possessions," and "we are in India to exercise Imperial sway . . . with the consent of the people and for their benefit": such were the arguments raised in public debate over the style of the buildings of New Delhi by supporters of an Indic style.[5] Broadminded and benevolent as it was meant to appear, regionalistic architecture that celebrated indigenous traditions, however, helped depict native cultures as tradition bound, demonstrating their need for guardianship from a modern European power.

The paternalistic stance inspired theories of government and methods of administration that also found expression in regionalistic architecture. One important corollary of colonial paternalism was the adoption of preservationistic policies aimed to "protect" indigenous culture from radical change; in other words, it was a trustee's duty to preserve and sustain, not radically reform. This often led to the adoption of "regionalistic" methods of administration, consciously adapted to the supposed horizons of native societies.[6] Although such paternalistic concern could sometimes reflect a genuine and deeply felt fascination, it could also harbor reactionary aspects. Thus, colonial development policy could amount to "development of underdevelopment," and its regionalistic methods could disguise social engineering. Occasionally the treatment of indigenous culture was outrightly manipulative. Regionalistic architecture could reflect all these aspects of colonial paternalism.

In regard to these issues, British sentiment toward Palestine seems to have been moved by two opposing tendencies: one assertive and possessive, the other paternalistic and empathic. In addition to imperial ambitions, the first was motivated by the Christian significance of the Holy Land and the memories of the Crusader quest; the second was typified by a romantic fascination with the Orient, perhaps best symbolized by the figure of Lawrence of Arabia in Bedouin costume. The staging of Allenby's famous entrance into Jerusalem on December 11, 1917, was, in fact, a rejection of the assertive-Crusader sentiment in favor of the paternalistic. What turned the British administration of Palestine decidedly away from a crusading pose was the concept of mandate, a legally formulated and internationally sanctioned realization of the doctrine of colonial trusteeship. In the covenant of the League of Nations, mandate status was literally defined as a "sacred trust of civilization." In Palestine such a mandate (instated in 1922, and finally ratified in 1923) justified the suppression of Crusader symbolism and assertive expressions of British dominance, and made strategies of colonial paternalism doubly appropriate. Before proceeding to examine some examples of regionalism in Mandatory architecture, it would be appropriate, however, to consider British attitudes toward Jerusalem, because it was in Jerusalem that most British buildings were erected, and it was the townscape of Jerusalem that these buildings primarily tried to portray.

JERUSALEM OF THE MANDATE

Making Jerusalem the capital of Mandatory Palestine and its administrative center, natural as the choice may seem, was a British decision. Focusing the country on a city where historic memory and traditional fabric had such an overwhelming presence had in itself preservationistic overtones. Something fundamental would have changed in the image of the Mandate had the focus of government been placed — plausibly enough — at the economic center of the country, the port town of Jaffa and by its fast-growing suburb of Tel Aviv.

Once Jerusalem was made the capital, its townscape and some of its major landmarks became all-Palestinian symbols. Thus, the Ottoman citadel and the Dome of the Rock were now featured on stamps, and the cubic, domed volumes characteristic of the domestic architecture of intramural Jerusalem and the Judaean vernacular became icons of the Palestinian landscape.[7] Such a fascination with the picturesque Oriental townscape had been largely absent during the Ottoman period. For example, the last Ottoman governor of Jerusalem, according to one source, had planned a tramway from the Jaffa Gate to the Haram which would have required extensive demolitions in the old city.[8] By contrast, Ronald Storrs, the British governor of the city from 1918–1926, swore (when the idea of a tramway to Bethlehem was raised) that a tram in Jerusalem would run over his dead body.[9] One of his first actions was to take measures to preserve the historic aspect of the city. He issued bylaws limiting any new development that threatened the historic townscape, commissioned town plans that defined areas for preservation, and created a voluntary Pro-Jerusalem Society that promoted the protection and restoration of monuments.

The European nineteenth-century heritage of Jerusalem, on the other hand, aroused much resentment. The monumental structures of the large religious and missionary-philanthropic institutions that competed for dominance in the landscape, and that announced their national affiliation through national architectural styles, were the antithesis of colonial paternalism: they were "interventionistic" in their purpose, as well as flagrantly "antiregionalistic" in their design.[10]

Likewise, the growing modern city aroused little enthusiasm. "Perhaps the thing that strikes one most," wrote one commentator in the 1930s, "is the change in relationship between the Old City within the walls, the Jerusalem of our hearts and minds, and the dreadful monster that has been spawned beside her."[11] In the introduction to a book commemorating planning in Mandatory Jerusalem, Allan Cunningham, the last high commissioner, perhaps best summed up the predominance of the old in British concerns in the city:

> The City of Jerusalem, precious as an emblem of several faiths, a site of spiritual beauty . . . has been in our care as a sacred trust for 30 years. In these pages will be found an important part of the story of the discharge of that trust, of the efforts made to conserve the old while adding the new in keeping with it, of marrying modern progress with treasured antiquity.[12]

"Marrying modern progress with treasured antiquity" was, in fact, a concise statement of the colonial approach to development, significantly associated here with the concept of trust.

ASHBEE AND HOLLIDAY, THE CIVIC ADVISORS

One of the first British architects to take part in the planning of the newly occupied Jerusalem was C.R. Ashbee, a leading figure of the Arts and Crafts movement, whom Ronald Storrs recruited in 1918 to act as a civic advisor.[13] Ashbee held this position until 1922, when he was replaced by Clifford Holliday. When Ashbee arrived in Jerusalem, ten years had already elapsed since he had lost the battle against industrialization and disbanded his famous "guild" in Chipping Campden. Medieval-looking Jerusalem was for Ashbee a heaven-sent opportunity to revive his notion of an Arts and Crafts utopia. In fact, Arts and Crafts ideals meshed perfectly with colonial preservationism. Characteristic of colonial observers, Ashbee was fascinated by traditional Palestinian society, and strove to protect its dignity and beauty from disruptive development:

> An industrialised Palestine . . . is a questionable benefit and, I think, a futile hope. Some new mechanical processes may be introduced . . . [but] may it not be better to leave as it is that peasant society, which still has so much dignity and beauty? . . . "New lamps for old" is an Eastern tale that still has value in the East, and Aladdin's mother, it is generally allowed, was a fond and foolish woman.[14]

Characteristically, Ashbee detested the "antiregionalistic" nineteenth-century European edifices of Jerusalem. "Coming into the City from the old pilgrim route, Mons Gaudii," he lamented, "we notice how the ancient Jerusalem is all but obliterated; we see the once golden dome no longer, we see a bastard Florence, a bastard Nuremberg, a bastard Moscow, an imitation Lourdes, a Bavarian suburb and an imitation Oxford."[15] The imperialistic meaning of these buildings did not escape him: "Round and about the City circuit within or without the walls we note an arrogant assertion of the various national codes. All modern buildings seem to have been strangely prophetic of the War."[16] The greatest sins of European penetration were, however, "commercialization" and "industrialization": "with the replacing of old workshop order by modern industrialism they have lost much of their ancient hold of life. In the old city they have run riot, and by mechanical means you can get infinite cheap and tawdry splendour."[17]

One of Ashbee's objectives in Jerusalem not only was to protect the old traditional fabric, but also to restore the authenticity of some of its picturesque sites. In the elegant volume published by the Pro-Jerusalem Society in 1921, he presented proposals for minute civic improvements, showing views "before" and "after" proposed intervention.[18] One proposal called for dismantling the corrugated iron shacks that had sprung up around the Jaffa Gate, and replacing them with structures built according to authentic, indigenous techniques, such that the "after" illustration depicted a picturesque arrangement of domed cubes, the archetypal image of the Palestinian vernacular. The new structures Ashbee envisaged near the Jaffa Gate, and near the Damascus Gate as well, were to be authentic, indigenous establishments: *khans* for the *felaheen* who came to sell their produce in Jerusalem and were obliged to find accommodation for the night for themselves and their camels **(Figure 4.1)**. Ashbee's sensitivity to local custom was at the same time somewhat reactionary. If *khans* were still in use in Palestine in 1918, the improvement of roads and transportation were soon to make them obsolete.

Yet even though none of Ashbee's *khans* was erected, the idea seems to have inspired at least one structure in Jerusalem, the Contagious Diseases Annex of the Ophthalmic Hospital of the Order of St. John. Unlike the Crusader order, its nineteenth century British revival was dedicated to providing medical care for Christian and infidel alike.[19] Since the nineteenth century, and up to the 1930s, the St. John Ophthalmic Hospital in Jerusalem had been the only specialized institution in the region where eye diseases could be treated. Villagers would travel miles on their camels and donkeys and camp all night around the gates in hope of being admitted in the morning when the hospital workday began. A traditional *khan* seemed the natural solution for this picturesque camp. According to Strathearn, the director of the hospital: "Those for whom rest accommodation is to be provided are native Mohammedans from outlying districts who are in the habit of sleeping in khans."[20] Thus, "it was intended that these cases should retain their own clothing and arrange for their feeding. They would be accommodated in rooms furnished with a 'mastaba' or low tiled platform around the walls and would sleep upon this, protected by a straw-mat or paillasse."[21]

However, the architectural office of Aston Webb & Son, who acted as architectural consultants of the St. John headquarters in London, commented, "It appears to us that if the Order are going to spend several thousand pounds on this building it should conform to modern hospital practice, with the necessary modifications."[22] In spite of the skepticism of the London office, the "regionalists" in Jerusalem had their way, and the *khan*, originally conceived in 1921 or 1922, was eventually designed by Holliday and consecrated in 1930. It consisted of a cloister-like court, surrounded by an open, vaulted arcade for visitors to sleep in. A fountain, set in an *iwan*, supplied water. The picturesque composition of the exterior was enlivened with Crusader motives and coats of arms. Characteristically, the photograph published in the order's publications, showed the structure with a figure in Oriental garb, posing at the gate.[23]

Equally offensive to Storrs' and Ashbee's tastes as the corrugated-iron encrustations around the city gates was the Ottoman clock tower erected over the Jaffa Gate in 1906.[24] Civic clock towers were introduced into the Ottoman townscape after 1839, when a small wooden tower was built near the Nusretia Mosque in Constantinople.[25] Abdülhamid II had had this structure replaced in 1888 by a permanent stone one and had added a number of other towers elsewhere in the city, including a particularly lavish one near the Europeanized Dolmabahçe Palace on the Bosphoros in 1894. Soon every town in the empire felt obliged to erect a civic clock tower. The fad arrived in Palestine during the period between 1901–1906.

Clock towers were conceived as symbols of Ottoman loyalty, but they also symbolized the new spirit of change that had spread through the empire since the

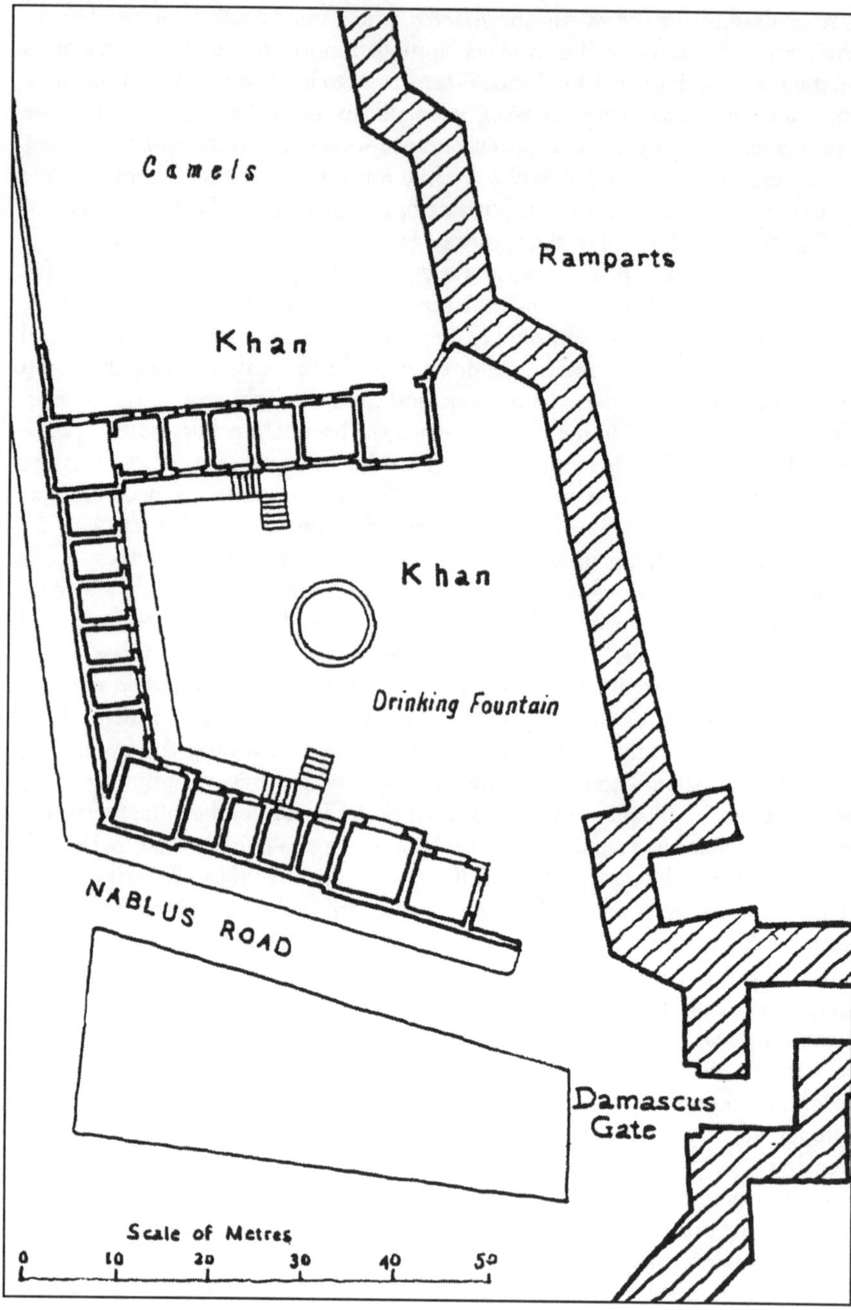

Figure 4.1. C.R. Ashbee, projected *khan* near Damascus Gate, Jerusalem, Plan. (Based on C.R. Ashbee, ed., *Jerusalem 1918–1920 being the Records of the Pro-Jerusalem Council during the Period of the British Military Administration*, 1921.)

mid-nineteenth century. Often built near newly erected municipal offices (local administration and municipalities were introduced only after the Ottoman municipal law of 1871 and 1877) and near newly formed city squares, the towers represented the emergence of a new sentiment of local patriotism and civic pride, as well as a new concept of public space. The mechanical clock introduced a new method of measuring time (hitherto associated with sun clocks in mosques), and therefore also signified a new perception of time. The architecture often represented a naïve mixture of European styles. A "style midway between that of Eddyson lighthouse and a jubilee memorial" is how Storrs described the one in Jerusalem.[26]

When Allenby marched into Jerusalem through the Jaffa Gate, the clock tower watched over the scene, a reminder of 400 years of Ottoman sovereignty. It was no wonder that the tower did not win British affection. More importantly, its naïve Westernizing style offended British regionalistic sentiment. Thus, in 1922 the Pro-Jerusalem Society managed, after some effort, to obtain the £500 necessary to dismantle the clock tower. But the story did not end here. When the intention to demolish the tower was announced, the city notables petitioned Storrs to spare the structure, which had been the pride of the city and had cost their community a considerable sum to erect ten years earlier. Storrs, who discovered that "a certain degree of sentiment attaches to the clock," promised that the tower would be re-erected elsewhere.[27] However, because the contractor did not manage to dismantle the tower without damaging the stones, the Pro-Jerusalem Society commissioned Holliday to design a new tower. The site chosen was an important intersection, Allenby Square (today Zahal Square), near the (since-demolished) central post office.

Holliday presented a model in January 1924, and by the end of the year, the structure was completed. It consisted of a series of shops (intended as a source of income for the society) arranged in an "L" form. The tower sprang from the angle of the L. According to the society, the design was in a style "appropriate for the city,"[28] that is, a subtle Orientalism: the shops had arcuated gates, and each was covered by a dome like a Palestinian house **(Figure 4.2)**. In contrast to the fussy Ottoman tower, the composition was based on simple volumes and smooth, white, stone surfaces. Thus, the Ottoman Victorianism of the old tower was replaced by British Orientalism, and the content of the civic symbol was reversed: although the inhabitants wished to adopt the trappings of Western progress, the British imposed on them the image of Orientals — improved and modernized perhaps — but Orientals all the same. The British clock tower was demolished in 1934 to make way for traffic.

HARRISON, CHIEF ARCHITECT OF THE PALESTINE PUBLIC WORKS DEPARTMENT

Austen St. Barbe Harrison (1893–1976), the Chief Architect of the Palestine Public Works Department (PWD), was the designer of all government buildings of Mandatory Palestine during 1922–1937, its architecturally most productive

Figure 4.2. Clifford Holliday, the clock tower at the end of Jaffa Road. Jerusalem, 1926. (Courtesy of the archives of the Centre for Architectural Heritage, Faculty of Architecture and Town Planning, Technion, Haifa.)

period. His large Palestinian *oeuvre* included two monumental representational buildings, the most prestigious projects undertaken by the Mandatory government: the High Commissioner's Residence, and the Palestine Archaeological Museum (now the Rockefeller Museum).[29] Harrison's deep identification with the spirit of the place made him a perfect author of a regionalistic architecture. Harrison dedicated a lifetime to design in the lands of the Eastern Mediterranean, and after finishing his studies in 1919, he returned to England only for short periods. When he began his fifteen-year career in Palestine, with hardly any practical experience, he had already adopted a regionalistic philosophy. A building must be "suited to its environment," "make sensible use of available materials," and "avoid flouting local tradition," he wrote in a memorandum concerning the design of the

High Commissioner's Residence in Jerusalem.[30] Faithful to his regionalistic convictions, he set out to study the forms and typology of Near Eastern architecture and the Palestinian vernacular, keeping abreast with the newest publications and archaeological discoveries; traveling extensively in Palestine, as well as in Egypt and Syria; and measuring and drawing buildings.

In his designs Harrison put this knowledge in practice in an original, and even radical way. Rather than merely add Oriental *caractère* to basically Beaux Arts plans, he took his inspiration from Islamic plan types. Thus, an early plan of the Archaeological Museum — Beaux Arts as it may seem — was also reminiscent of the Faraj and Barquq mausoleum in Cairo.[31] Harrison's plan for the residence of the Chief British Representative in Amman was based on a four-*iwan* plan, the same arrangement found in the Çinili Kiosk at the Topkapisaray in Istanbul **(Figure 4.3).**[32] The drawing room at the High Commissioner's Residence in Jerusalem, the centerpiece of Harrison's design for this structure, is also based on the four-*iwan* plan.[33] And in an unrealized project for a British Council for Jerusalem (1942), he designed an inner court dominated by a towering single *iwan*, no doubt drawn from Cairene *madrasas*.[34]

Harrison's use of the four-*iwan* plan in the residences at Amman and Jerusalem was perhaps the most remarkable. As he must have known, this plan type was associated in Islamic domestic architecture with audience halls and imperial *divans*. Providing a British representative with such a hall implicitly presented him as an Oriental ruler. In this regard, the gesture was somewhat similar to the provision of a Durbar Hall, a hall for traditional Mughal imperial audiences, in the Viceroy's Palace in New Delhi — although with Harrison the architectural reference was perhaps meant as subtle metaphor, possibly tongue-in-cheek.

For specifically Palestinian character, Harrison turned to the vaulted construction of the Judaean vernacular. "There is a particular kind of rubble vaulting peculiar to this country," he explained in a report, "which not long before the War was everywhere employed and today is still employed outside the cities. This vaulting,

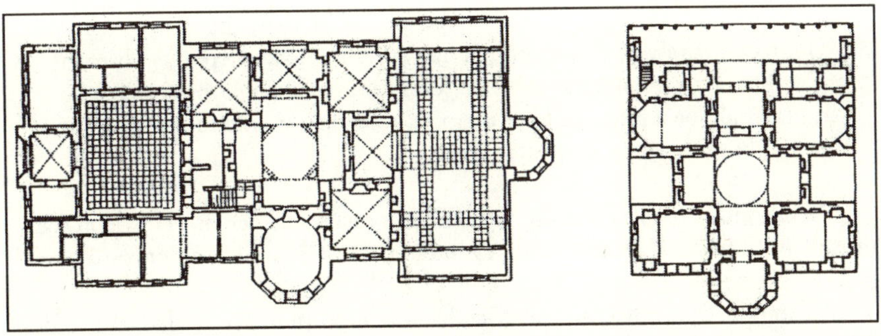

Figure 4.3. Austen St. Barbe Harrison, the Residence of the British Representative, Amman, plan; and the plan of its likely source of inspiration, Çinili Kiosk, Istanbul. (Drawings by authors.)

merely whitewashed or painted, is so pleasing to the eye that any kind of decoration or 'finish' becomes superfluous."[35] Harrison borrowed the aesthetics as well as the construction of the vernacular vault in all his representational buildings. Thus, the residences in Amman and Jerusalem and the Archaeological Museum all have imposing vaulted halls constructed in the indigenous method. Photographs record local Arab masons in traditional costume working on Harrison's building sites. Letting local Arab builders erect the vaulted structures in the traditional techniques not only ensured utmost authenticity, but also served to prove that Palestine was still truly traditional. Modern methods were still unfeasible in Palestine, Harrison told a reporter: "Stone was used, not for sentiment, but because it was found that local workmen were so well acquainted with old methods. . . . To have taught a large corps of laborers the modern use of warped concrete surfaces would [not] only have taken considerable time and money, but would have made less beautiful results."[36] Nevertheless, concrete was manufactured locally after 1925, and the less prestigious spaces were constructed of reinforced concrete.

Some of Harrison's buildings in Jerusalem required more than a general response to the spirit of the place. In the case of the High Commissioner's Residence and the Palestine Archaeological Museum, the sites selected for the buildings called for consideration of the physical context of Jerusalem itself — its fabric and townscape. The museum occupied a particularly sensitive site overlooking the northeast corner of the walled city. Climbing the path from the Jericho Road to the museum gates led the visitor along the imposing city walls, promising a view of Jerusalem's roofscape as the road rose higher. A full view of the city was to be offered from a "map room" containing maps and models of the city at the top of an "outlook tower."[37] Such a scheme would have made the Old City itself into an exhibit, perhaps the prize exhibit of the museum; but at the same time, it would have interfered with its skyline. A tall tower in the immediate proximity of the revered Old City also would have repeated the sins of the hated nineteenth century establishments. At least this was the opinion of the Director of Antiquities, E.T. Richmond:

> So striking an architectural feature as a tower would give the building the appearance of dominating, from those points of view, the ancient City. This would surely be unfortunate; it might even be looked upon as a breach of architectural good manners. . . . Jerusalem has enough towers already; . . . it would be a pity to add to them, or to appear to join in competition for unnecessarily attracting attention.[38]

Richmond's "preservationist" stance here nears self-denial. Harrison, who himself was "divided" on this point, eventually was prepared to compromise, and lowered the tower.

Although the museum practically touched the city walls, the High Commissioner's Residence enjoyed a breathtaking panorama of the Old City from a well-removed vantage point on the south. There was an implied appropriation in this. What an impressive gesture it must have been for the high commissioner to point out to a visitor all the famous landmarks and holy sites seen from the house. At the same time, the residence was too far from the center of town to dominate its skyline. In contrast with the assertive nineteenth century edifices, it seemed to watch over the city from afar, as if wishing merely to supervise — to "hold the ring," rather than dominate. Furthermore, although offering a perfect view of old Jerusalem, it gave no view of the new city. Nothing reinforced the traditional, historic aspect of Palestine in the mind of the British administration more than the view from the High Commissioner's Residence.

Underlining only the reactionary, preservationistic implications of Harrison's Orientalism does not fully represent the message of his architecture. Unlike Ashbee, Harrison did make room for modernity in his vision of Palestine. In the spirit of colonial regionalism, however, he was looking for a modernity that evolved from tradition and that agreed with the spirit of the place: "Marry modern progress with treasured antiquity" to use Cunningham's expression **(Figure 4.4)**. This ambition was not entirely colonial: the dichotomies of tradition versus modernity and regionalism versus universalism preoccupied British and European architects between the wars. Unlike the Modern Movement, which sought a new start for architecture altogether, traditionalists wished to achieve modernity without breaking with the past. The Italian architect Marcello Piacentini concisely expressed this ambition in a review of an article by Le Corbusier in *L'Esprit Nouveau* in 1921: the task at hand, he wrote, "involves, basically, resolving the debate between impersonal, international, standardized architecture, and localized vernacular architecture."[39]

Figure 4.4. Austen St. Barbe Harrison, the High Commissioner's Residence, Jerusalem, 1931. (Courtesy of G.S. Schweig and the archives of the Centre for Architectural Heritage, Faculty of Architecture and Town Planning, Technion, Haifa.)

Piacentini also indicated a traditionalist's solution for this conflict: "but then are the two really antithetical? . . . it is possible to arrive at a vision of a sane architecture which will be neither old nor new, but simply true."[40] Piacentini was advocating what may be termed the transcendental strategy: by following "simply true" principles of architecture (i.e., principles that were perceived as universally and eternally valid), the conflict between the international and the localized, the old and the new, could be resolved — or more accurately, transcended. By underlining the universal and eternal values of past tradition, the old could be proved equally modern, whereas the nagging problems of the here and now could be presented as petty and ephemeral.

For many architects the universal and immutable principles of architecture were those stated in the theoretical tradition and design discipline of Classicism as embodied in Beaux Arts teachings. Thus, British architects like Lutyens (in his late, Palladian phase) and Charles Holden shared, with other European architects of the 1910s and 1920s, a search for a radical Classicism that transcended style and reached to the rudiments of architecture. In the colonial context, this strategy was a perfect solution to the dilemmas of colonial representation. By focusing on the most abstract and general precepts of Classicism — such as rational composition, geometrical ordering, primary forms — European design could be made valid for the Orient. Likewise, by abstracting indigenous tradition of its relative, specific aspects, it could be universalized. The whole range of colonial conflicts — regionalism versus universalism, modernity versus tradition, East versus West, imperialistic versus paternalistic — could thus be resolved, or again, rather, transcended. In New Delhi, Lutyens' "elemental manner," combined with his subtle abstraction of Oriental forms, suggested this strategy.[41] Arthur Shoosmith, one of his disciples in New Delhi, carried it to the extreme. In his St. Martin's Garrison Church (1927), Shoosmith distilled his master's "elemental manner" of all vestiges of Classical detail to achieve an abstract composition of masses that subtly alluded to some unspecified, primitive and ageless vernacular tradition. In this way he managed to evoke indigenousness and, at the same time, speak a universal language of geometry — to be British, and at the same time, of India.[42]

In his monumental Palestinian buildings, Harrison, Shoosmith's exact contemporary, pursued his own interpretation of the transcendental strategy. His use of axially ordered Islamic/Near-Eastern plan types seems to demonstrate how well Oriental compositions agreed with rational Beaux Arts procedures. The four-*iwan* plan, with its two major center lines, could fit perfectly with an axial Beaux Arts composition. Furthermore, when abstracted of detail, Harrison seemed to imply, Çinili Kiosk — the four-*iwan* plan that inspired the plans of the residences at Amman and Jerusalem — was not so far apart from a Western archetypal pavilion, such as the Villa Rotunda (or its English adaptation, Chiswick House). The cross-into-square articulation of the tower at the High Commissioner's Residence employs a common turn-of-the-century mannerism that apparently derives from a Villa Rotunda–like composition.[43]

Superimposing it over the cruciform arrangement of the four-*iwan* drawing room commented on the affinity between an Oriental and Classical archetype.

Although the universal ideal of rational composition helped Harrison to disprove the discord between Oriental and Classical design, he used radical abstraction to universalize the Oriental formal vocabulary and make it compatible with Beaux Arts conceptions. Thus, Harrison avoided any Classical reference, but instead drew his inspiration eclectically from a wide range of Near Eastern precedents. These, however, remained inspiration only. Rather than simply cite them, Harrison extracted from his precedents a limited number of primary forms, such as the round arch and the pointed arch, from which he rationally and methodically developed a diverse typology of architectural elements — arcades, doorways, window moldings, and niches. These were vaguely Oriental, but at the same time could be read as elementary compositions independent of precedent. All was designed in emphatic simplicity, the aura of Platonic perfection completed by the use of glaringly white local limestone.

"Near-Eastern Modernism" is how a critic characterized Harrison's universalizing treatment of Oriental and vernacular sources.[44] Although aspiring to transcend time and place, his approach was nevertheless characteristic of progressive British architecture of his time. Thus, Harrison's abstraction of historical precedent amounted to a sort of "stripped Orientalism," comparable to the stripped Classicism explored at the time by Holden and others; and the stark monumentality of the white limestone masses recalled the Portland-stone monumentality then in vogue in Britain. At the same time, this ponderous interwar style could harmonize perfectly with the landscape of Palestine. The sharp Mediterranean sun gave Harrison's purism a vitality that its Portland-stone counterparts rarely enjoyed under British skies.

Of his designs, the Archaeological Museum best demonstrated how Harrison was able to pursue the universalizing — and, to his mind, also modern — reading of the Palestinian vernacular to achieve an imaginative and moving architectural portrait of Jerusalem. His point of departure was the domed cube, the icon of the Judaean vernacular and the basic element of the townscape of Jerusalem. On this archetypal element he based a typology of turrets, barbicans, pavilions, rotundas and the one large tower. These were all variations, ranging from the simple to the elaborate, that derived from the basic prismatic form of the domed cube. Harrison inserted these as nodal accents in various crucial points of the plan. When viewed from a distance, especially from the south, this cluster of forms seems to echo the silhouette of the Old City. In fact, the whole complex appears like a small city, a rationalized reflection of the real Jerusalem. The cluster of white prismatic forms that tower over the city from the hill call to mind Le Corbusier's famous vision of Rome, with a row of primary bodies hovering above it. It is an image of Jerusalem transformed through the power of reason into a heavenly city where the contradictions and irrationalities of its earthly counterpart are resolved. It was a sublime portrait: so sublime, in fact, that it lost sight of its living counterpart and its pressing modern problems.

GEDDES, MEARS, AND THE HEBREW UNIVERSITY ON MOUNT SCOPUS

The Mandate of Palestine included an important and unprecedented stipulation, upon which representational buildings would be expected to comment: making Palestine into a Jewish "national home." The popular movement for Jewish national revival in Palestine began in Eastern Europe in the late nineteenth century, and quickly became a worldwide, well-organized political movement. Modern Jewish settlement in Palestine began in 1882, and by the time of World War I had become a significant entity. It already boasted a network of thriving agricultural settlements and fast-growing urban quarters around Jaffa (Tel Aviv), Jerusalem and Haifa. A new Jewish-Palestinian culture was also emerging: Hebrew-speaking, dynamic, assertive and aggressively modernist. The British pledge to establish in Palestine a "Jewish national home" (Balfour Declaration) was incorporated in the wording of the Mandate, and thus the British government became internationally committed to the continued prosperity of the Yishuv. Yet, because the Mandate also stipulated that the rights of Palestinian Arabs should not be compromised, the whole history of the British administration in Palestine revolved around attempts to reconcile the conflicting national aspirations of the two communities. But British difficulties with the Jewish national home had a more general source than such a conflict of nationalisms: the very idea of a Jewish national home conflicted with Britain's paternalistic colonial vision. In fact, the idea of a Jewish national home proved the antithesis of colonial preservationism: although colonial philosophy believed in tradition, regionalism, and "slow and sound progress," the Yishuv represented modernism, cosmopolitanism and aggressive development. Worst of all, such a movement threatened to disrupt the image of the tradition-bound, timeless, Oriental Palestine that colonial paternalism was geared to protect. Reconciling the Jewish national home with Arab Palestine was yet another aspect of such basic colonial dilemmas as assertiveness versus paternalism, and interventionism versus preservationism, faced by the British administration.

Representational Mandatory buildings such as the High Commissioner's Residence and the Palestine Archaeological Museum did not make any explicit acknowledgment of the Jewish aspect of the Mandate. Therefore, it was left to the Yishuv to search for its own architectural symbolism. Strangely enough, the Zionist Organization chose a British planner, Patrick Geddes, to design the grandest of all Jewish projects in Palestine, the Hebrew University on Mount Scopus.[45] For ten years (1918–1928), Geddes, with the help of his son-in-law, the architect Frank C. Mears, struggled to realize their vision of the university, the ultimate Jewish cultural symbol in Palestine. Eventually, only a small portion of Geddes' and Mears' plan, the National Library (today the Law Faculty of the Hebrew University) was ever realized.[46]

Patrick Geddes is known as a diverse and visionary thinker, a pioneer of British modern town planning. Because a holistic worldview was fundamental to his thinking, he approached every issue with a defiant disregard for polarities, seeking sweeping, world-embracing continuities. Thus, he envisaged human experience as a many-celled matrix that charted the worlds of action and spirit, of reality and thought, and systematically covered all the possible interactions between the different categories of human existence. For the planner, this holistic view dictated an awareness of the continuities of time and place, which, in turn, spelled out evolutionistic and regionalistic solutions. It is not common today to think of Geddes as a colonial planner, but, in fact, he dedicated most of his planning career to the colonial context (primarily India and Palestine). When applied to the colonial context, his approach strangely appears as a rather exalted and longwinded reiteration of the conceptions of colonial preservationism:

> While the Western (and engineer-educated) sanitarian [or planner] . . . complacently applies his English and Victorian manufacturing town experience to Indian sanitary problems . . . this complacent assurance is impossible to the planner who is anything of a geographer and anthropologist. For he sees the people of different climates and environments as adapted through past ages to these. Thus he comes to their ways, their habits, their customs, their institutions, their laws, their morals, their manners, with ordinary naturalistic attitude of observant and interpretative interests, and not that of superiority.[47]

Zionism, as a movement that sought renewal of an ancient culture through a return to its "authentic habitat," held a special appeal to Geddes. It was a dramatic realization of his own theories on the new emerging from the ancient and on restoring harmony between culture and place. Furthermore, Zionism symbolized for Geddes the revival of the universal values that he seemed to find in Judaism. Particularly attractive to Geddes was the holistic ideal implied by monotheism. Thus, Zionism, as Geddes saw it, promised no less than to reform universal culture on the lines of his own convictions: "reviving Jerusalem stands for . . . a true re-Hebraizing of Europe and the West; that is renewal of ancient discernment of Unity, and this not only in the cosmic universe but in the moral also."[48] Geddes then proceeded to compare Jerusalem to Athens (where reconstruction projects were being contemplated at the time, 1919, after the Balkan wars):

> That these two main culture-cities of the past . . . should thus in the present . . . be rekindling their ancient altars, is thus more significant than even they, much less the world, yet see. For they must each renew their ancient message to the world. . . . Though to Helvetize the world be thus the great aim of the League of Nations, the success of this cannot but be in the measure in which the world is at once re-Hellenized, re-Hebraized as well. Re-Christianized thus also, for Christianity is the union of both elements. . . . The future will quarrel less about the names of religions; their essential value is the measure of their reuniting spirit of love.[49]

This sublimated, universalized Zionism could therefore agree without any difficulty with Palestinian Arab culture. And once the universal, humanistic meaning of Zionism was revealed, Jews could live in harmony with their Muslim and Christian Arab brothers: "The fears of unending strife between Jew, Arab and Christian over Palestine do not exist for me. There is never any permanent need for people to kill each other. And for all three religions, there is unity from past origins."[50] Looking back to the past, following the continuity of human history to its origins, Geddes sought to discover the common sources that would ensure present harmony. And because any cultural process grows out of a harmonious interaction with its physical context, he determined that "old Palestine" must be the germ for any renewal in the country: "It is thus out of the old Palestine, with its broken and surviving beauty, that men of good will, each and every faith, may best hope to express their own vision of their new Jerusalem upon its hills."[51] Through this happy turn of argument, indigenous Palestinian culture offered Geddes a program for a modern architectural idiom for Palestine:

> Look at any good photograph of an Arab hillside village. . . . [S]ee the bright walls and brighter dome-roofs in the sunshine, and how, as it were, they chime together. . . . These simple houses and small domes make up the essential picture, ranging from sunrise joy to sunset glory; and thus they justify the bigger dome that here and there give them value. Here is architecture in its very essence. . . . This old-world architecture — which seems to have been Jewish before it was Arab, if not as old as the conquest of Joshua — expresses the historical life of the land, and is thus fully capable of development with its modern life and spirit.[52]

In other words, although Zionism could be universalized to make it compatible with Arab Palestine, the presumed timelessness of the Palestinian vernacular could offer inspiration for Zionist representational architecture.

The Hebrew University offered Geddes an opportunity to realize his ideas on a grand scale. In fact, Geddes' belief in grand continuities induced him to exceed his commission and offer proposals that affected the whole city, and indeed the whole of Palestine. Because the university was to be a part of the life of the city, its departments had to be involved in the culture, education and economy of the city. He conceived, therefore, of a whole array of institutions, not only placed on Mt. Scopus, but also in different appropriate sites in Jerusalem. Besides academic institutes, these included an experimental school, medical school, a series of museums, housing for staff and students, and a garden-city for the employees of high-tech industries connected with the university.[53]

The plan of the Mt. Scopus campus was a graphic realization of Geddes' charts of human experience. It consisted of an octopus-like complex, each tentacle (six in number) representing a different branch of knowledge. In the hub of the complex, there was a large hexagonal square, in the center of which stood a gigantic domed hall, the "Dome of Synthesis," as Geddes called it **(Figure 4.5)**. It was meant to house facilities for the daily social life of the university and a center for academic

Figure 4.5. Auila Academia of Geddes's Hebrew University Project. Image from a souvenir program of May 21, 1921. (Courtesy of Department of Manuscripts and Archives, Jewish National University Library, Jerusalem, File V.1145, postcard in sleeve.)

conventions ("Aula Academica"). On the ground floor, Geddes planned gymnastic facilities, and in the galleries he envisaged exhibitions depicting the history and life of the city. Because the western side of the hexagonal square was to be left free of structures, a sweeping panorama of the old city would unfold from the hall. The hall, therefore, represented a perfect Geddesian synthesis: the coming together of all the branches of science, the interaction of spirit and body, between scholarship and daily life, and between the university and the city. It was also an extravagant exercise in *architecture parlant*, overlaid with the symbolism of Geddes' private mythology. The dome was, in fact, a temple for Geddes' own philosophy.

The dome, however, also carried Jewish symbolism. The hexagonal geometry of the court suggested the form of the Star of David, reproduced by Geddes and Mears on the interior of the dome. The star, and above all, the dome itself were for Geddes the symbols of unity, the great contribution of Judaism to the world. In his words, the Great Dome represented "the great Unity which it is the glory of Israel to have first realized and taught, and which Christian and Moslem alike, despite all their hostility, have the candor to confess they have thus inherited."[54] In a letter Geddes suggested that he gave the dome the same diameter as the Dome of the Rock, a magnificent view of which opened from the site. He must have known that the Dome of the Rock modeled its dimensions on the Rotunda of the Anastasis at the Holy Sepulcher. Three large domes of the same diameter, each representing one of the three great religions, would have been an invincible proof of their unity. Yet Geddes' enthusiasm for unity carried him even further, and engulfed the religious symbolism of any dome possible:

> The dome is of universal use for religious expression — not only common alike to the Pantheon of old Rome, to the church and mosque of Santa Sophia of Constantinople, of St. Peter's and St. Paul's of Catholic and Protestant, to all the sacred buildings of Islam, but also to the great religions of the further Orient.[55]

It is amazing that Geddes claimed to have found the inspiration for this world-embracing dome symbolism in the archetypal domed cube of the Palestinian vernacular. But the idea hit him when he inspected the site on Mount Scopus. "Looking more carefully at the landscape . . . of the hill range, . . . is not the very best of all the buildings upon the whole sky-line the single old Mosque, with its small domes and very modest minaret upon the southern of the two heights of the Mount of Olives?"[56] Carried away by his enthusiasm for domes, Geddes now suggested to echo the Great Dome by a multitude of secondary domes. The whole ridge from Mt. Scopus to the Mount of Olives was to bubble with domes, thus paraphrasing the domed houses with the occasional "bigger dome that here and there give them value" of the old city. Even housing for industrial workers was depicted in the perspective drawings as domed cubes — the Arab village turned into a garden city. Geddes and Mears must have occasionally felt that deriving Jewish symbolism from Palestinian Arab architecture was a somewhat problematic idea, but they thought they managed to solve the problem: "To distinguish and develop the Jewish style from the Arab or Saracenic is of course a great problem — and I think we are succeeding."[57] The beautiful perspective drawings were the proof: "Well look at the plans and see! — Neo-Palestinian are they not? — instead of Neo- (or Paleo-) European."[58]

Geddes was a great supporter of Zionism, but the Zionism he supported was a Zionism sublimated and reinterpreted to fit his own conceptions. To resolve the conflict between Zionism and Arab Palestine he sought first to universalize it, and then "regionalize" it. The re-Hebraization of the world ended thus in the Orientalization of Zionism. Independently of the colonial administration, Geddes adopted a paternalistic and regionalistic reasoning that in spite of its exalted tone was nevertheless in the spirit of colonial regionalism. Uniquely, however, it was applied to the Yishuv, not to native Palestinian society.[59]

CONCLUSION

The image of Jerusalem was the chief preoccupation of the British architects whose work we have surveyed here. The townscape of the intramural city was almost an object of worship: architects reveled in offering a view of it, gave much thought to the impact of their buildings on it, and, above all, tried to portray and characterize it in their designs. It was the Oriental aspect of the city and its characteristic vernacular architecture that British architects celebrated. To an extent, the Oriental and traditional representation of Jerusalem corresponded to the British conception of the whole of Palestine.

The preoccupation with the spirit of the place and the fascination with traditional native culture was inevitably colored by a colonial perspective. The examples we have explored suggest all the shades of what we have termed "colonial regionalism," that is, regionalism motivated by a paternalistic conception of the role of the colonial power.

The Orientalism of the High Commissioner's Residence announced the paternalistic nature of the British rule most directly. By celebrating the Oriental and the indigenous — rather than conveying Britishness — in a building that by its very purpose symbolized British authority, the residence depicted the high commissioner as a ruler of Palestine acting for the benefit of the country, rather than a deputy of the British crown looking after the interests of the empire.

Preservationism, the wish to protect indigenous culture from change, is most blatantly represented in the case of the city clock tower. Whereas the Ottoman tower had been conceived as a symbol of modernity, the British tower spoke of the persistence of a native, Oriental culture. Similarly, Ashbee's proposed *khans* were not merely symbolic gestures, but an attempt actually to preserve a traditional way of life. This was regionalism in a functional rather than aesthetic sense, a conscious attempt to adapt planning to indigenous manners and customs. By seeking to sustain an already-anachronistic institution, his idea was as reactionary as it was patronizing, however. Aston Webb & Son's criticism of the choice to adopt the form of a *khan* for the St. John Hospital extension made explicit the conflict between regionalism and modernity.

Harrison's regionalism — unlike Ashbee's — represented an ambition to balance tradition and modernity. Harrison hoped to resolve, or rather transcend, the basic contradictions of colonial regionalism by adhering to principles he believed to be universally and eternally valid. In the world of absolute values — or so his "abstract Orientalism" seemed to suggest — there was no meaning to contradictions between East and West, indigenous and universal, traditional and modern. Although this strategy muted the poignancy of colonial paternalism and preservationism, it made for a somewhat detached and irrelevant result, just as did Shoosmith's St. Martin's Church in New Delhi.

The manipulative aspect of colonial regionalism was not absent in Mandatory architecture. Perhaps the most fundamental point is the fact that British regionalistic architecture did not merely depict Palestine, but helped to invent a tradition for a newly created state. When British architects selected icons of Palestinian culture to inspire their designs, not only did they make a comment on the nature of Palestinian identity, but took part in defining it. Buildings such as the High Commissioner's Residence and the Archaeological Museum were elaborate national symbols, devised by the good services of the British authorities for the benefit of Palestine.

Nevertheless, British commitment to establishing in Palestine a Jewish national home contradicted colonial preservationist conceptions. Conflicting Jewish and Palestinian Arab national aspirations introduced a new dimension to the hybrid urbanism of Jerusalem. Geddes' design for the Hebrew University employed regionalism to attempt to reconcile Zionism to Palestine, as well as make it conform with the conceptions of colonial paternalism. This was a case of cultural manipulation, applied however, to Zionism.

Thus, in spite of the uniqueness of Palestine, and the special meaning of Jerusalem, it seems that the conceptions that guided its architectural representation basically derived from the colonial situation. The cultural, ethnic, religious and historic complexity of Jerusalem was eventually subjected to a colonial critique. Thus, the regionalistic line of British Mandatory architecture puts it in the same category as some of the British architecture in India, and British treatment of Jerusalem recalls in some respects the case of Rabat under the French protectorate.

However, realizing the colonial content of British architecture in Jerusalem does not fully sum up its meaning. Each of the architects discussed here was guided, independently of the colonial theories, by his own architectural convictions: Ashbee by Arts and Crafts ideas, Harrison by Beaux Arts theories and the strategies of interwar traditionalism, and Geddes by his own world-embracing holistic philosophy. It was the colonial context that brought out the paternalistic and preservationistic element implied in these convictions. Their work in Jerusalem represented, therefore, an interesting hybridization between British architectural theories and colonial thinking, unexpectedly applied to the complex reality of the Holy Land.

The colonial nature of Mandatory architecture should not mean that it possessed nothing to commend it. Much British design in Palestine is aesthetically and intellectually compelling. Few would totally disprove of some measures for the conservation of the intramural city and for the protection of its townscape, objectives to which the British administration made important contributions. More importantly, some of the dilemmas that preoccupied designers like Harrison and Geddes — if not their solutions — are universal. Such difficulties as reconciling modernity and tradition, the indigenous and the global, and balancing national aspirations with universal humanistic values will always be relevant issues. The British representation of Jerusalem perhaps did not offer a lasting formula for shaping the identity of a deeply contested and ambiguous place, but it added yet another layer of meaning to the complex historical connotations of the holy city.

NOTES

This chapter is based on material from R. Fuchs, "Austen St. Barbe Harrison: A British Architect in the Holy Land," D.Sc. dissertation submitted to Technion, Haifa, 1992 (in Hebrew).

1. On the historical geography and architecture of Ottoman Jerusalem, see Y. Ben-Arieh, *Jerusalem in the 19th Century: The Old City*, Jerusaelm, Yad Ben-Zvi, 1984; *Jerusalem in the 19th Century: Emergence of the New City*, Jerusalem, Yad Ben-Zvi, 1986; D. Kroyanker, *Jerusalem Architecture*, Steimatzky, Jerusalem, 1994; R. Kark and M. Oren-Nordhein, *Jerusalem and Its Environs 1800–1948*, Jerusalem, 1995 (in Hebrew, English version forthcoming); and R. Hillenbrand (ed.), *Ottoman Jerusalem*, Fox, London, 1998. On specific aspects, see R. Kark and Sh. Landman, "The Establishment of Muslim

Neighbourhoods in Jerusalem Outside the Old City During the Late Ottoman Period," *Palestine Exploration Quarterly*, 1980/81, vol.112/113, pp.113–135; and M. Crinson, *Empire Building, Orientalism and Victorian Architecture*, London, Routledge, 1996, pp.198–226. On Jerusalem during the period of the British Mandate, see B. Hyman, "British Planners in Palestine 1918–1936," Ph.D. diss., London School of Economics and Political Science, 1994; D.B. Monk, "An Aesthetic Occupation: Architecture, Politics and the Menace of Monuments in Mandate-Era Palestine, 1917–1929," Ph.D. diss., Princeton University, 1995; and R. Kark and M. Oren-Nordheim, "Colonial Cities in Palestine? Jerusalem under the British Mandate," *Israel Affairs*, vol.3, no.2, 1996, pp.50–94.

2. British architecture in Mandatory Palestine is largely the creation of half a dozen British architects. In addition to Austen St. Barbe Harrison, these included Clifford Holliday, Charles Robert Ashbee, Patrick Geddes, Frank Mears, John Burnet, Henry Kendall, and Benjamin Chaikin. For basic information on these architects, see the relevant entries in *The Dictionary of Art*, London, Macmillan, 1996.

3. F. Béguin et al., *Arabisances, limits et grandes tracés*, Paris, 1978; F. Béguin, *Arabisances: decor architectural et tracé urbain en Afrique du nord, 1830–1950*, Paris, Dunod, 1983; G. Baudez and F. Béguin, "Arabisances," *Lotus International*, vol.26, 1980, pp.41–52; Th.R. Metcalf, *An Imperial Vision: Indian Architecture and Britain's Raj*, Berkeley, University of California Press, 1989; P. Rabinow, *French Modern*, Cambridge, MIT Press, 1989; G. Wright, *The Politic of Design in French Colonial Urbanism*, Chicago, University of Chicago Press, 1991; N. AlSayyad (ed.), *Forms of Dominance*, Aldershot, Avebury, 1992; P.R. Baudel et al. (eds.), "Figures de l'orientalisme en architecture," *Revue du Monde Musulman et la Méditerranée*, vol.73/74, 1994, pp.3–4; and M. Given, "Star of the Parthenon, Cypriot Melange: Education and Representation in Colonial Cyprus," *Journal of Mediterranean Studies*, vol.7, no.1, pp.59–82.

4. Edmund Burke in a speech at Parliament in a debate over Fox's East India Bill, 1783.

5. Quoted from Ernest B. Havell, in Metcalf, *Imperial Vision*, p.213, n.4; and from John Begg, in Ibid., p.216, n.10. References dating from 1912 and 1913.

6. For a concise account of "preservationism" and "regionalism" in colonial government, as well as opposing tendencies, see the chapter on "Social Engineering" in D.A. Low, *Lion Rampant*, London, Frank Cass, 1973. See also P. Hetherington, *British Paternalism and Africa 1920–1940*, London, Frank Cass, 1978.

7. For a recent account of the Palestinian vernacular, see R. Fuchs, "The Palestinian Arab House and the Islamic 'Primitive Hut,'" *Muqarnas*, vol.15, 1998, pp.157–177.

8. C. Holliday, "Town and Country Planning in Palestine," *Palestine & Middle East Economic Magazine*, no.7/8, 1933, p.290.

9. Cited, e.g., in H.B. Samuel, *Unholy Memories of the Holy Land*, London, Woolf, 1930, pp.199–200

10. E.g., St. George College (begun after 1890) drew on New College and Magdalen College, Oxford; the Russian Compound (the 1860s onward) and the Russian church at Gathsemane cited typical Russian mannerisms including onion domes; the Augusta Victoria (1906–1910) and the Dormition Church (1900–1910), were based on German Romanesque precedents; and finally the Tuscan-style Italian Hospital (1913–1919). On European building activity in Jerusalem, see Ben-Arieh, *The New City*; Crinson, *Empire Building*; and E. Meyer, "Die Dormition auf dem Berge Zion in Jerusalem," *Architectura* 14, 1984, pp.149–170.

11. D. Duff, *Palestine Picture*, London, Hodder & Soughton, 1936, p.57.

12. H. Kendall, *Jerusalem, the City Plan, Preservation and Development during the British Mandate*, HMSO, London, 1948, p.v.

13. On Ashbee, including a chapter on his work in Jerusalem, see A. Crawford, *C.R. Ashbee: Architect, Designer & Romantic Socialist*, New Haven, Yale University Press, 1985.

14. C.R. Ashbee, *Palestine Notebook*, London, Heinemann, 1923, p.270.

15. C.R. Ashbee, "Report on the Arts and Crafts of Jerusalem and District, 1918," typescript, possession of Felicity Ashbee, photocopy in the Jerusalem City Archives, article 128.

16. Ibid., article 130.

17. Ibid., article 127.

18. C.R. Ashbee (ed.), *Jerusalem 1918–1920 being the Records of the Pro-Jerusalem Council during the Period of the British Military Administration*, 1921. See also C.R. Ashbee (ed.), *Jerusalem 1920–1922 being the Records of the Pro-Jerusalem Council during the First Two Years of the Civil Administration*, 1924.

19. On the order, see E. Bradford, *The Shield and the Sword: The Knights of St. John*, Hodder & Stoughton, London, 1972. On the activities of the revived order in Jerusalem since the nineteenth century, see E.J. King, *The Knights of St. John in the British Empire*, London, St. John's Gate, 1934, pp.163–174; E. Hume, *Medical Work of the Knights Hospitallers of Saint John of Jerusalem*, Baltimore, Johns Hopkins, 1940, pp.299–346; and Ben-Arieh, *The New City*, p.324.

20. A report by Arthur E. Franklin, April 1925, St. John Archives, St. John's Gate, London.

21. The Warden, Strathearn, to the Hon. Secretary, W. Perowne, St. John's Gate, January 7, 1925, London, in ibid.

22. Maurice Webb of Aston Webb & Son, to Perowne, March 2, 1925, in Ibid.

23. King, *The Knights of St. John*, opposite p.169.

24. On the Jerusalem clock tower, see Ben-Arieh, *The Old City*, pp.31–37; B. St. Laurent and A. Riedlmayer, "Restorations of Jerusalem and the Dome of the Rock and Their Political Significance, 1537–1928," *Muqarnas*, vol.10, 1993, pp.82–83.

25. On the Ottoman clock towers, see G. Goodwin, *A History of Ottoman Architecture*, London, Thames & Hudson, 1971, pp.380,419; and Z. Çelik, *The Remaking of Istanbul*, University of Washington Press, 1986, pp.130–131.

26. Ashbee, *Jerusalem 1918–1920*, introduction, p.vi.

27. Storrs to Rouhi Bey, February 23, 1922, Protocols of the Pro-Jerusalem Society, Jerusalem City Archive.

28. Protocol no.50 of the Pro-Jerusalem Society, January 25, 1924.

29. On Harrison, see Obituary, *Times*, February 14, 1976; a short biographical sketch in the chapter on Nuffield College in H. Colvin, *Unbuilt Oxford*, New Haven, Yale University Press, 1983; A. Suszmann and R. Reich, "The History of the Rockefeller Museum," in E. Schiller (ed.), *Zeev Vilnai Book*, vol.2, Jerusalem, 1987, pp.83–91 (in Hebrew); Fuchs, "Austen St. Barbe Harrison," 1992; and *The Dictionary of Art*, Macmillan, London, 1996.

30. Harrison to Pudsey, Dir. PWD, September 21, 1926, Israel State Archives (ISA) 3/3/3(1), doc.4127

31. PWD Archives, Jerusalem, reproduced in Fuchs, "Austen St. Barbe Harrison." Harrison may have seen the plan in M.S. Briggs, *Muhammadan Architecture in Egypt and Palestine*, Oxford, Clarendon, 1924.

32. See plan in "The British Residence, Amman, Transjordan," *The Architects' Journal*, December 28, 1932, p.835. The plan may also have been inspired by the early

Islamic four-*iwan* pavilion on the acropolis at Amman. The affinity between the Çinili Kiosk and the Amman pavilion was pointed out in publications Harrison may have noted: W. Bachmann et al., *Petra*, Berlin, Gruyter, 1921, fig.62, p.71.

33. Plans and photographs in Ch. Hussey, "A Crusader Castle of Today," *Country Life*, vol.LXX, no.1815, October 1931, pp.480–486.

34. Reproduced in D. Kroyanker, *Jerusalem Architecture: The Period of the British Mandate 1918–1948*, Jerusalem, Keter, 1989 (in Hebrew), p.122.

35. Harrison to Dir. of PWD, Jerusalem, Memorandum on the Bethlehem Road project for the High Commissioner's Residence, August 24, 1927, CO733/145/57021/1A. On Palestinian vaulting, see Fuchs, "The Palestinian Arab House," pp.167–168.

36. H. Taylor, "Palestine's New Museum," *New York Times*, August 10, 1936.

37. "A Builder of Jerusalem, Austen Harrison," an unidentified newspaper clipping in Harrison's archive, Israel Museum, Jerusalem. The projected map room was never carried out.

38. January 21, 1928, ATQ 202, Israel Authority of Antiquities Archives, Jerusalem.

39. "Le Corbusier's 'The Engineer's Aesthetic,'" *Architettura e Arti Decorative*, II, 1922, pp.220–223, translated in P. Serenyi (ed.), *Le Corbusier in Perspective*, Englewood Cliffs, NJ, Princeton-Hall, 1975, pp.26–27.

40. Ibid. Other architects made similar arguments. See, especially, reference to Howard Robertson's publications in A. Colquhoun, *Modernity and the Classical Tradition*, Cambridge, MA, MIT Press, 1989, pp.45–46.

41. On Lutyens' New Delhi work, see the relevant passages in Ch. Hussey, *The Life of Edwin Lutyens*, London, Country Life, 1950, reprinted 1984; R.G. Irving, *Indian Summer*, New Haven, Yale University Press, 1981; and Metcalf, *An Imperial Vision*.

42. On Shoosmith and the St. Martin's Garrison Church, see G. Stamp, "Indian Summer," *Architectural Review*, no.159, June 1976, pp.365–372; and Irving, *Indian Summer*, pp.334–338.

43. Compare, e.g., Holden's Incorporated Law Society, London 1903–1904, Lutyens' Midland Bank, Manchester 1929; F.L. Wright, Unity Temple, 1906; and J. Hoffmann, Maison Stoclet, 1905.

44. Hussey, "A Crusader Castle."

45. For recent studies on the history of the university, see Sh. Katz and M. Heyd (eds.), *The History of the Hebrew University of Jerusalem, Origins and Beginning*, Jerusalem, Magnes, 1997 (in Hebrew).

46. On Geddes' and Mears' design, see the relevant passages in A. Defries, *The Interpreter Geddes*, New York, Boni & Liveright, 1928; Ph. Boardman, *Patrick Geddes Maker of the Future*, University of North Carolina Press, Chapel Hill, 1944; *The Worlds of Patrick Geddes*, London, 1978; H. Meller, *Patrick Geddes, Social Evolutionist and City Planner*, London, Routledge & Paul Kegan, 1990; Fuchs, "Austen St. Barbe Harrison"; Hyman, "British Planners in Palestine"; V.M. Welter, "The Geddes Vision of the Region as City: Palestine as 'Polis,'" in J. Fiedler (ed.), *Social Utopias of the Twenties*, Wuppertal; Müller & Busmann, 1995, pp.72–79; M. Shapiro, "The University and the City: Patrick Geddes and the First Master Plan of the Hebrew University, 1919," in Katz and Heyd (eds.), *The History of the Hebrew University*, pp.201–256; D. Dolev, "The Hebrew University Master Plans, 1918–1948," in ibid., pp.257–280; and "Architectural Orientalism in the Hebrew University: The Patrick Geddes and Frank Mears Master-Plan, *Assaph-Studies in Art History*, Tel Aviv University, no.3, 1998, pp.217–234.

47. Geddes, *Indore Report*, vol.1, 1918, p.73, quoted in Meller, *Patrick Geddes*, p.265.

48. Defries, *The Interpreter Geddes*, pp.268–690.

49. Ibid.

50. Ibid., p.262.

51. P. Geddes, "Palestine in Renewal," *The Contemporary Review*, 670, October 1921, p.478.

52. Ibid., p.477.

53. For a detailed account of Geddes' proposals, see Hyman, "British Planners in Palestine"; or Shapiro, "The University and the City."

54. Geddes to Stephan Wise, July 18, 1929, Mss.275–290, National Library Scotland. I am grateful to B. Hayman for letting me read his microfilmed copies.

55. "Geddes: Proposed University of Jerusalem," *The Scotsman*, May 18, 1921.

56. Geddes and Mears, "The Proposed Hebrew University of Jerusalem," December 1919, stenciled typescript, p.26, copy Central Zionist Archives L4/108H.

57. Geddes to Miss Macleod, November 11, 1919, Mss.275–290, National Library Scotland.

58. Geddes to Eder, November 13, 1920, in ibid.

59. A universal, spiritual conception of Zionism was not foreign to Zionist thought, including to the approach of such leading figures in the history of the University as J. Magnes and H. Weizmann. The interaction between the conceptions of British planners and those of the Yishuv, as well as Palestinian Arabs, is a subject for another essay. For the Palestinian perspective on the evolution of Palestinian identity, see R. Khalidi, *Palestinian Identity: The Construction of Modern National Consciousness*, New York, Columbia University Press, 1997.

PART II

MODERNITY, GLOBALIZATION AND URBAN FORM

5

Stages of Globalization in the African Context: Mombasa

Ali A. Mazrui

Mombasa is Kenya's second city. To that extent it may be worth considering whether second cities generally deserve a different theoretical framework. Is there a need to study second cities as special entities? Is Montreal a second city to Toronto? Is Melbourne a second city to Sydney? Should scholars approach the subject in a different manner?

Mombasa is also a kind of global city — although it is small as global cities go. It is the second largest city not only in Kenya, but in East Africa. One is still talking about a city of less than a million people. Nevertheless, it is the largest seaport of the region, and it is one of the oldest urban centers in East Africa. It has been around for quite a while — long enough that the nature of its globalization has had three fairly distinct and different phases. Within each there have also been subphases or stages.

First one must bear in mind that Africa as a whole is culturally intermediate between Europe and Asia. There are parts of Africa that are more "Occidental" than any part of Asia. Then there are parts of North Africa and parts of South Africa that are more Europeanized than one can find anywhere on the Asian continent. On the African continent, one may find bits of Europe which are not reproduced anywhere in Asia. On the other hand, there are parts of Africa which have more in common with Asia than have any parts of Europe. Islam, for example, is older in parts of Africa than it is in most of Asia. And Christianity is older in Ethiopia than it is in most of Western Europe. This essential intermediate position which Africa enjoys between the Occident and the Orient is captured in the story of Mombasa. It is part of the story of this African city on the Indian Ocean.

As indicated, there have been three distinct phases in the history of Mombasa as a city. The Afro-Oriental phase was the period when Mombasa was a small cultural arena where traditions from West Asia and South Asia interacted with

African traditions in search of new cultural configurations. This period covered several hundred years, during most of which Mombasa was known as Mvita.

The second historical period was the Afro-Occidental phase when Mombasa was at last "discovered" by Europeans. First the Portuguese and later the British initiated the process of reorienting Mombasa away from its traditional links with Asia, and toward a new relationship with Europeans and with Western culture.

The third phase of Mombasa's history is currently in progress. This is the Afroglobal phase, during which the city has come to feel the pull of both East and West. The old transition from being a city-state to being the main port of a nation-state has now greater global repercussions for Mombasa and its people.

Let me describe each of these phases of Mombasa's globalization in turn.

THE AFRO-ORIENTAL PHASE

Three interrelated factors conditioned the origins of Mombasa's Afro-Oriental phase. One factor was that the city lay in the path of the monsoon winds, and was therefore reachable by trading ships sailing from Asia.[1] Secondly, Mombasa itself had a natural harbor which could accommodate those ships, as well as activate a nautical and shipbuilding tradition of its own. Thirdly, Mombasa was relatively near the Arabian peninsula and was soon to interact culturally, as well as commercially, with the land that gave birth to the Muslim religion. Mombasa became one of Islam's entry points into eastern Africa.

As a result of this configuration of factors, the multiculturalization of Mombasa began quite early. Consumption patterns felt the cultural influences which came with the trading traffic of the monsoons. From Arabia and other ports of Asia came expensive porcelain and plates, brass-decorated wooden chests, silks and finery.[2] In East Africa these were sometimes paid for in ivory, copra and animal skins. There was also a limited trade in slaves.

At the more popular level of consumption culture, the Afro-Oriental phase of Mombasa's history produced the increasing use of seasoning and spices in the local cuisine. New spices were imported into Mombasa from Asia, and innovative, syncretic and eclectic cuisines developed among the coastal people generally, combining elements of African, Indian and Arab food cultures.

It was during the Afro-Oriental phase that rice became the staple diet of the people of Mombasa and surrounding coast areas. Even though the vocabulary of ordinary rice dishes remained completely African (*mpunga* for unhusked rice, *mchele* for uncooked rice, and *wali* for cooked rice), dishes for special occasions employed such borrowed words as *biriyani* and *pilau* from South Asia. On the other hand, the names of the individual spices used in preparing these quasi-Indian dishes were more likely to have been borrowed from Arabic than from any of the Indian languages (*bizari* for curry powder and *thumu* for garlic, for example).

The Afro-Oriental phase also witnessed changes in the musical culture of Mombasa and the coastal areas. In addition to the traditional African drums, the small Indian drum (*tabla*) found its way into the coastal range of musical instruments. In time, the Arabian string instrument, the *ud*, also entered Swahili culture. So did the dancing drums with small bells attached known as *matari*. Indeed, in some Islamic societies along the coast *matari* were even used in mosques as accompaniment to certain lively praise songs in honor of the Prophet Mohammed. (More traditionalist Muslim theologians strongly disapproved of the use of drums in the mosque or for any religious purpose.[3])

The flute in coastal music was a child of all three traditions — African, Arab and Indian — allowing each to reinforce the other, but the "Orientalization" of Mombasa music was the dominant trend.[4] Over time, even the tunes of Swahili songs became strongly influenced by Arab (especially Egyptian) music, on the one side, and Indian music on the other. At its worst, some so-called Swahili composers simply plagiarized the music of Indian songs and substituted Swahili lyrics. Fortunately, there always remained a hard-core Swahili tradition which tried to remain authentic against all comers from the Orient.[5]

The most profound changes which came from Asia into Mombasa concerned religion and language. Islam arrived in eastern Africa while the Prophet Mohammed was still alive, but Islam's first African landing space was not Mombasa but a part of Ethiopia. Persecuted Muslims from Arabia arrived in Abyssinia in quest of asylum during the Prophet of Islam's own lifetime. It is because of this that some have argued that the *hijra* (migration for asylum) from Arabia to Africa was almost as early, and in a few respects almost as symbolic, as the Prophet's own *hijra* from Mecca to Medina. With the African *hijra*, a seed was planted which, by the end of the twentieth century, had turned Africa into the first continent with an absolute Muslim majority.

Not long after the death of the Prophet Mohammed, Islam reached the part of eastern Africa that now includes Mombasa. This meant that mosques were built in East Africa before they were constructed in parts of what is now the Middle East. Islam in Mombasa is older than Islam in Istanbul and the rest of Turkey. It may be older than Islam in Islamabad and the rest of Pakistan. Is it older than Islam in parts of what is now the Arab world? Certainly parts of the Maghreb in North Africa (such as Morocco) were probably penetrated by Islam later than the area of Mombasa, though this remains a matter of historical calculation.

Before long, the arrival of Islam in Mombasa and surrounding areas had a diverse effect on the cultural experience of the people.[6] Marriage and kinship relations were changed profoundly, as were the rules of inheritance and succession. African indigenous norms often came into competition with the new Islamic rules. In some cases syncretism was the result; in others the indigenous norms maintained their edge; but increasingly the Afro-Oriental phase of Mombasa's history witnessed the gradual triumph of Islamic rules of marriage, kinship, inheritance and succession.

The arrival of Islam in Mombasa also had a profound impact on dress culture. The concept of nakedness was completely redefined for both men and women, with practical consequences for forms of attire for each gender. The *kanzu* entered the scene for men (the long outer garment). Ultimately, it became religion neutral in Uganda, where both Christians and Muslims accepted it as a kind of national dress. In Kenya, however, the *kanzu* remained associated only with Muslims.

In Mombasa the womenfolk developed the black *buibui* for outdoor use, not only intended to veil their faces but also to deny shape to their bodies. The *buibui* was worn by Muslim women of Mombasa only out of doors when they visited relatives or went grocery shopping. The shapelessness of the garment was part of Islam's quest for female modesty in public places. "In public do not emphasize the curves! Conscience begins with avoidance of temptation!" This is a strong Islamic premise. It profoundly affected dress culture in Mombasa during the Afro-Oriental phase.

The arrival of Islam in Mombasa also affected architecture. In general, such influence had begun with the introduction of construction methods using coral masonry and lime plaster by early traders arriving from the region of the Red Sea. Then, as masonry construction techniques were refined, Kenyan coastal buildings began to take on aspects of design common around the western Indian Ocean. Thus developed an architectural tradition of flat-terraced roofs, carefully articulated arched doorways and refined plaster surfaces.

The arrival of Islam also shaped East African coastal cities such as Mombasa in more programmatic ways. The mosque became the dominant public institution and architectural focus of the urban community. However, its plan and style developed distinctively, built without the towering minarets or grand, central courtyards of Asian and Arabian models. Mosques of Mombasa featured fully enclosed prayer halls and *mihrabs* made of multiple rebated arches, often decorated with inset porcelain bowls. Following a popular tradition found throughout the Islamic world, tombs of departed Islamic holy men also became the focus of a ritual of monumental construction. Large palace structures were built with an elegance and on a scale that allowed them to serve as symbols of power and provide monumental settings for public audiences and rituals.

The homes of the ordinary residents of Mombasa also felt the influence of Islam. Domestic structures were increasingly designed around formal courtyards, following notions of privacy and gender segregation common to Islamic cultures. Such residences also contained places of prayer for women with attached ablution rooms. At times their innermost walls were decorated with neatly constructed niches.

Following a long period in which these influences were paramount, the winds of the Western world began to be felt in Mombasa. East was East, and West was West — but the twain were to meet along the East African coast in Mombasa.

THE AFRO-OCCIDENTAL PHASE

In the process of the Westernization or Occidentalization which began at the beginning of the sixteenth century in Mombasa, there was always one Western power that was dominant. Again, three phases are historically identifiable. First was the phase when Portugal was predominant, virtually unchallenged in the region. Then came the more comprehensive phase of European colonialism when the British had a controlling presence in Kenya as a whole, and in Mombasa. Thirdly, there is now the postcolonial phase, during which the major Western influence in the region is that of the United States. Mombasa has felt each successive shift in power and influence.

Vasco da Gama was virtually the first European to disembark in Mombasa when he arrived in April 1498. The sailors of Mombasa and other settlements along the East African coast had initially helped him in his efforts to find his way to Calicut in India. Although the people of Mombasa did not realize it at the time, the Portuguese explorer was also preparing the ground for his nation's imperial incursion into eastern Africa. The Portuguese later came with warships and sought to build coastal African colonies to service their trading fleets in the Orient. In Mombasa they built Fort Jesus, a fortress to defend and monopolize the old harbor, and from which they sought to rule Mombasa and its environment.

How was Mombasa "Westernized" in this first phase of the Afro-Occidental era? Architecture was partly influenced by the Portuguese, as some Portuguese homes were emulated by the locals. The Portuguese also used military architecture to dominate the local population **(Figure 5.1)**. The Portuguese language also

Figure 5.1. A remnant of a Portuguese fortification at the southern end of the island facing out to sea. It has now been modified for use as a rest stop and billboard. (Courtesy of Thomas Gensheimer.)

had an impact on local languages. Thus, such present-day Swahili words as those for slippers (*sapatu*), female undergarment (*shimizi*), and a certain water kettle (*kindirinya*) are originally derived from Portuguese. The old Mombasa dialect even used the metaphor *mreno* (Portuguese person) to mean any kind of aristocrat.

Perhaps the most enduring cultural impact of the Portuguese was on the food culture of Mombasa, and subsequently of East Africa as a whole. Because of Portugal's links with the so-called New World of the Americas, the Portuguese in East Africa eventually brought maize and potatoes, which assumed increasing importance in the diet of the local people. In time, many of the indigenous people of the coast (such as the Giriama) adopted *ugali*, or maize-meal (powdered dry maize), as their staple food — a trend which subsequently spread to the hinterland of Kenya. The Waswahili of Mombasa continued to use rice as the staple, but they increased the use of potatoes as a supportive dietary element and as a feature of minor daily meals. Maize on the cob and *ugali* were also used for variation.

The Portuguese impact on the cuisine of Mombasa did not endure, except perhaps in the case of some desserts and puddings — an influence that was later reinforced by the British.[7]

Westernization under the Portuguese also meant the Westernization of warfare for the first time in Mombasa. The use of gunpowder, cannons, and guns brought a new phase in the city's history. Later on, this influence would prove a "nemesis" for the Portuguese, as local people eventually used Western-style weapons to drive them out of Mombasa. The Mazrui family played an important role in combating the Portuguese and finally driving them out of their fortress, Fort Jesus. And after the Mazrui family occupied the fort, they ruled the Mombasa city-state from it for about 100 years, beginning at about the time of the French Revolution of 1789.[8]

Clearly the most far-reaching period of Westernization in Mombasa began in the late nineteenth century, when British colonial rule was established **(Figure 5.2)**. This phase continued until December 1963, when the independent modern nation of Kenya was born. British colonial policy in Mombasa was far more comprehensive than was Portuguese. It included the building of primary and secondary schools for local children, the establishment of a legal system for the society as a whole, the demarcation of Kenya as a future nation-state, and its inclusion into an elaborate system of accountability within a global British Empire.

Within every 24 hours, the children of Mombasa might cross civilizations several times. They might speak Swahili (or Kiswahili) at home, take Arabic or Islamic lessons at the mosque, watch Indian films on weekends, and be forced to speak English at school.

Of these several civilizations under the British Raj, there was no doubt which eventually became the most ascendant in Mombasa. Western civilization exerted increasing influence at school, at the Kilindini harbor, in the new films at local cinemas, and as a direct result of the influence of British power.

In particular, the English language was implanted to a degree that Portuguese never had been during Portugal's occupation of the city. A school

Figure 5.2. Historical view of the Mombasa skyline at the beginning of the British colonial era. (Published in many texts. Originally from Sir Richard Burton, *Zanzibar: City, Island and Coast*, vol.2, 1872.)

syllabus for British-style education was established, linked to an empire-wide system based on examinations set by the University of Cambridge in England, and school teachers included increasing numbers of men and women from the United Kingdom sent out to serve the British Empire. Of course, white missionaries were also active, seeking to spread the Christian gospel through missionary schools, church services, and medical clinics bearing the cross.

The British not only helped spread literacy in English but also in some of the indigenous languages of what would become Kenya. In Mombasa the relevant indigenous language was Swahili (or Kiswahili). In the first half of the twentieth century, newspapers began to appear in the city in both languages. Of these, the *Mombasa Times*, published daily in English, became the most influential. It was edited and run by Europeans. Other racial groups in Mombasa (African, Arab and Indian) developed weekly or monthly publications of their own in diverse languages. The languages were diverse, but the tradition of newspapers and weekly periodicals in any language reflected a process of Westernization.

Under British rule, however, Kenya also became a white-settler colony. British colonial administrators encouraged large-scale white immigration to farm the fertile soil of what was overtly called "the white highlands." Although these highlands were several hundred miles from Mombasa, the transformation of Kenya into a white-settler colony had profound implications for the city.

Although Mombasa was an ancient city, it was ultimately bypassed to build a new British capital closer to the "white highlands." Nairobi was rapidly developed from a minor train stop on the Uganda railway to a major African metropolis. By the second half of the twentieth century, Nairobi had outstripped Mombasa in size. What saved Mombasa from shrinking into insignificance altogether was its continuing strategic value as the most important port in East Africa, regardless of where the Kenya capital was.

During the years of British rule, Mombasa and other parts of Kenya were influenced by Western tastes in cultural terms as well. Dress culture among the younger generation borrowed more and more from the West. Manners, etiquette, and even forms of boasting became influenced by Western norms. As John Plamenatz, the late Oxford philosopher, once put it: "The vices of the strong acquire some of the prestige of strength."[9]

Above all, the colonial mentality of the time looked to the imperial metropole for a general standard of life and culture. Caliban sought the guidance of Prospero. The British way of life was for awhile the ultimate standard of civilization to the colonized. Mombasa fell into the shadow of this mental dependency.

Such emulation of British ways, however, eventually also came to include British standards of democratic life. Thus, when the undemocratic nature of colonial rule became broadly evident, nationalists in Mombasa and other parts of Kenya started demanding "One Man, One Vote." British standards of democracy by the second half of the twentieth century meant universal adult suffrage. This could only lead to full independence. In mid-December 1963, Mombasa became the proud port of independent Kenya. Old-style colonialism was over.

It was during the postcolonial period following Kenyan independence that the focus of Westernization in Mombasa shifted from Britain to America. The shift was neither sharp nor total; after all, Britain and the United States were themselves close allies, and both were English-speaking powers. Nevertheless, in the postcolonial world capitalism mattered more than colonial ties, and the leader of the capitalist world was obviously the United States.

American influence in determining the prices of Kenya's agricultural products (especially coffee and tea), American influence with the World Bank and the International Monetary Fund, and American influence among the community of aid-donors generally contributed to this shift from Britannic to Americo-centric Westernization. Meanwhile, expanding trade, aid, and investment relations between the United States and Kenya also directly strengthened ties between the two countries.

American influence was also felt through the increasing impact of American music, television programs, films, and magazines, which were making major cultural inroads into Mombasa and Nairobi. Young people also responded to the American dress culture — with special reference to jeans and T-shirts. And the American genius in fast food affected the Mombasa street and restaurant culture. Thus did the hamburger arrive on the African side of the Indian Ocean, as well as Kentucky Fried Chicken, or its equivalent. Competing with African, Arab and Indian snacks of old, there were now new fast mini-meals from the food culture of the United States. In Mombasa, these were particularly appealing to young people. The United States was clearly the epicenter of this new burst of postcolonial Westernization.

It has today been an easy transition from the Westernizing phase of Mombasa's historical experience (the Afro-Occidental phase) to the Afro-globalist phase. It is to this even more wide-ranging stage of Mombasa's historical evolution I now turn.

THE AFRO-GLOBAL PHASE

Three forces have contributed to the globalization of Mombasa: war, tourism, and international politics. The wars have ranged from World War II to the Mau Mau insurrection; tourism's impact has been felt since the days of Theodore Roosevelt; and the effect of international politics has included the repercussions of the Cold War.

Of course, there were earlier stages in the globalization of Mombasa; and the first two phases in the city's history (the Oriental and the Occidental) were themselves manifestations of global forces. But there were always other, more local, trends evident during those times.

The present era may be said to have begun when Mombasa's port acquired special significance during World War II. As a seaport, Mombasa was important both in the defeat of Italy early in the war and in transporting African troops to Burma in the later fight against Japan. Mombasa was also important to the flow of goods between Africa and Europe all throughout the wartime period.

It is ironic that the ancient name of Mombasa was Mvita, meaning Isle of War. The name derived from the fact the city was fought over often because of its strategic location. In the twentieth century, Mombasa once again became the Isle of War, but in a vastly different sense. It has been estimated that about 500,000 African soldiers from British colonies served in World War II. They not only fought in the Horn of Africa and in North Africa but also in Italy and in the Axis-occupied parts of the Fertile Crescent. Africa's heaviest casualties were in the war in Burma against the Japanese. A large proportion of those African soldiers sailed from Mombasa.

Almost none of the African soldiers who fought in World War II were allowed to rise above the rank of sergeant, and they were almost always led by white officers. Although the contributions of Australians, New Zealanders and other Commonwealth combatants have repeatedly been celebrated (including during the 1994 D-Day extravaganza), very little tribute has ever been paid to the African heroes and martyrs of World War II. Mombasa was to many of these African soldiers what Portsmouth was to the soldiers who invaded Normandy on June 6, 1944: a point of departure on a momentous journey.

When Italy's East African colonies of Ethiopia and Eritrea were overrun by the British, Mombasa was used to accommodate many Italian prisoners of war, both men and women. Some of the earliest interracial dating experienced by Mombasa's local African population took place between Mombasa men and Italian women detained there during World War II. Before World War II, sex between blacks and whites in Kenya had been a crime, so interracial dating during the war must itself be counted as a stage in the city's globalization. Curiously enough interracial dating also involved Polish refugees temporarily encamped in Mombasa.

Later on, Mombasa acquired new military value for the British in relation to the Suez Canal. After World War II, Egyptian nationalists put increasing pressure on Britain to vacate its military base near the canal. After the Egyptian revolution of 1952, these nationalist pressures became irresistible. One result was that Mombasa, in combination with Aden in Yemen, acquired additional significance in British strategic calculations. The two ports constituted the fall-back position after the loss of the Egyptian military prize.

As part of these events, Mombasa found itself absorbed into a wider circle of strategic calculations. Unfortunately for British planners, the year of the Egyptian revolution was also the year of the outbreak of the Mau Mau war in Kenya. The twin events unleashed forces that eventually resulted in the eviction of the British from both countries, and the British were subjected to the same militant fate in Aden.

The Mau Mau war in Kenya was itself a globalizing experience, giving the colonial politics of Kenya much more publicity than they had ever had before. As part of this struggle, the name Jomo Kenyatta became known to households across the world. Although the war was fought mainly in central Kenya rather than at the coast, it did, of course, affect Mombasa. For one, the British used Mombasa to bring in troops, planes and ammunition. In the end, the British military victory over the Mau Mau nationalists proved illusory. Although the Mau Mau were defeated, their political cause of Black rule in Kenya prevailed.

The next stage in the military history of Mombasa followed the election of Jimmy Carter as president of the United States in 1976. It was then that the American concept of a "rapid-deployment force" in defense of the oil resources of the Middle East was developed. This required a cluster of seaports which could be used by U.S. ships either for refueling or "morale-boosting" holidays for servicemen and -women charged with guarding the approaches to the Red Sea and the Gulf. Before long, the beginnings of an American military base began to take shape in Mombasa. All through the 1980s, American naval ships stopped periodically in Mombasa, and American servicemen disembarked. The city's shops, nightclubs, hawkers and whorehouses soon learned the skills of catering to visiting American servicemen. Mombasa was absorbed into the more militarized side of the Cold War.

In a sense, tourism as a globalizing experience in Kenya is older than war. In the first half of the twentieth century, Mombasa served as the main point of entry for all overseas tourists coming to East Africa. But the coast of Kenya was not the major attraction of the country for European and North American tourists at that time. This was the safari — traveling into the bush, tracking wild animals either to photograph or kill. These were the days of animal trophies — the most ambitious being leopard skins, elephant tusks, or the head of a lion.

The white hunters came through Mombasa, and left with their trophies from Mombasa. One of the most famous tourists in Kenya at this time was Theodore Roosevelt, the 26th president of the U.S. from 1901 to 1909.

Although safaris and wild beasts continue to be a major part of Kenya's tourist package, by the 1980s hunting in Kenya had been outlawed. By this time old-style safari hunters only represented a tiny percentage of tourists in any case. In touristic terms, Mombasa has now evolved into more than a simple point of entry for tourists. This has largely been due to the city's own distinctive cultural mixture and to the development of the beaches along the Kenyan coast. Some of the major beach hotels have been built in Mombasa — as have expensive beach chalets for longer-term rental.

Beachfront development has not been without its own controversies. What may first appear to be positive regional economic development often results from projects put together by outside business interests. These developments may displace local people, who then suddenly find themselves as squatters with no land rights. Insult is added to injury when newly created jobs are reserved for nonlocal groups.

Thus, with the coming of the jet age, although Mombasa's importance as a tourist point of entry has been drastically reduced, its significance as a tourist destination has risen dramatically. In cultural terms, one must ask whether such touristic globalization has not neutralized the city's previous Afro-Orientalization. It is true Westernization has taken place partly at the expense of both the city's African and Oriental ways. The new globalization is undoubtedly an ally of Westernization, and both continue to erode the city's African and Oriental legacies. In some cases, coastal tourist projects have even taken the ubiquitous form of modern high-rise buildings (with four-star accommodations), providing considerable impetus to the globalization of local building forms.

As the city has learned to cater to Western tourists and American service personnel, the pleasures of the evening have also become Westernized. More nightclubs, dance bars, and Western-style restaurants have opened. And the old rhythms of African drums, the Arab *ud*, and the Indian *tabla* and *sitar* have given way to the aggressive guitars and drums of the Western world. Indeed, African music itself has become substantially Westernized in the last twenty years.

One cruel irony in this situation today concerns fears generated by Acquired Immune Deficiency Syndrome (AIDS). Trends that might otherwise have led to a comprehensive Westernization of Mombasa may have been slowed by these fears. Certainly the sexual liberalization that came with Westernization has become more restrained in the 1990s. Tourists arriving in Mombasa seeking "white-sand beaches, the blue ocean, and black sex" have become much more careful about this third component. So have their former black partners, who regard white folks as the original carriers of AIDS into Africa.

It is one of the ironies of history that cultural Westernization in Mombasa is being slowed partly because of the fear of AIDS. The African and Oriental cultures of Mombasa may be the accidental beneficiaries of a terrible disease.

The Bible used historical floods and famines to tell a moral story. One day historians may use AIDS to draw definitive cultural conclusions.

When did international politics begin to influence the globalization of Mombasa? Such a process probably began during the previous phase of Mombasa's historical experience. Certainly, the anti-colonial struggle in Kenya was an experience in international politics. During this time the name of Tom Mboya, the most prominent Kenyan politician in the 1950s after Jomo Kenyatta, became well known in the United States and Great Britain. Mboya lectured widely in defense of the Black nationalist cause in Kenya, and was an eloquent illustration of how internationalism contributed to globalization.

Even his assassination in 1969 in Nairobi was a form of internationalization — although his assassins were local. This was because he was an important African friend of Martin Luther King Jr., who had himself been assassinated. In the early 1960s, I had met Dr. King in New York City. I already knew Tom Mboya personally at the time. Ironically my only conversation with Martin Luther King had in fact been, in part, about Tom Mboya. Before the end of the decade both Mboya and King had fallen to an assassin's bullet.

In the 1950s Mombasa was important in Mboya's rise to national prominence. He was a trade union leader who became involved in a vital dock strike in Mombasa of the 1950s. His combined skills of aggression and negotiation propelled him to national visibility. It was from within the ranks of trade unionism that he then entered wider anti-colonial politics.

After independence, the Cold War also became a major factor in the globalization of Mombasa. Although Jomo Kenyatta had been denounced by the British as "leader into darkness and death" and had suffered as a Mau Mau detainee, he emerged from prison ready to forgive and forget. He became one of Kenya's leading Anglophiles and took the country firmly into the capitalist camp in the Cold War.

However, Mombasa's port at the time served a larger geographical area than just Kenya. Uganda under Milton Obote was attempting to "Move to the Left" ideologically, and this had a tense effect on relations with Kenya. Uganda needed Mombasa to export its goods and import goods from outside. Poor relations with Kenya sometimes tempted Uganda to explore alternative points of access to the sea. But, unfortunately for Uganda, Mombasa was still by far the best outlet to the sea.

Under Idi Amin, Uganda's domestic policy became capitalist, but its foreign policy became increasingly anti-Western and pro-Soviet. Amin's administration also had problems with the Kenyan government — which sometimes led to Kenyan threats to withhold the use of port services at Mombasa. Here again a paradox emerged. Mombasa's role as an asset for all of East Africa was sometimes compromised by regional and global politics. Thus, the Cold War cast its shadow on Mombasa directly and through the inter-state rivalries of East Africa.

The globalization of Mombasa also occurred through the United Nations. Of its agencies, the United Nations High Commissioner for Refugees (UNHCR) has needed Mombasa to deal with the problem of refugees from Somalia. It has even established refugee camps in the city. And the United

Nations' Environmental Program (UNEP) has established its global headquarters in Nairobi. As the first UN agency to have its headquarters in Africa, UNEP has naturally become closely involved in issues related to its immediate environment. This has included the debate about pollution in the Indian Ocean. How culpable is Mombasa in relation to the pollution of the sea?

Africa's internationalization has also meant inter-Africanization. During World War II, Mombasa experienced what were perhaps its first visitors from West Africa — the soldiers from the Gold Coast (now Ghana). Mombasa also saw uniformed white South Africans on their way to battle areas. The inter-Africanization of East Africa has reached new levels since the 1940s. Because Kenya has become one of the leading diplomatic and conference centers of Africa, Mombasa has inevitably been a pan-African beneficiary.

Paradoxically, Afro-globalization has also included a partial return to Afro-Orientalization. Islam in Mombasa in the 1990s became more politically radicalized, echoing tendencies toward Islamic militancy elsewhere in the world.

But no city exists independently of the state and nation to which it belongs. Although the city has gone through its three stages towards globalization, Kenya, as a state, has had a somewhat different transition. The national and state dilemmas have had implications for Mombasa. It is to this wider context of Kenya and its orientation that I now turn.

THE NATIONAL CONTEXT: BETWEEN THE ATLANTIC AND THE INDIAN OCEANS

Visitors rarely enter Mombasa today as they were intended to: by sea. Rather, they slip in surreptitiously from the back, entering along the narrow isthmus of land that barely connects the island city to the mainland. From this perspective it is difficult to get a sense of the city; it appears at first like some neglected provincial capital, rather than a thriving commercial port. Yet, the streets leading to the city center are lined with rows of thriving shops and commercial structures — looking like one large urban market. The scene provides a continuing link to the city's past and its traditional source of wealth. Mombasa's roots as an international center are also revealed in its diversity of building types, including Hindu temples and schools, modern mosques built in the Indian style, arcaded buildings and broad streets left by the British, and modern storefronts which represent the functional language of commerce.

Such a thriving commercial center coexists uneasily with its past. Prosperity has shifted away from the original urban center, the "Old Town," which now appears largely as a curiosity rather than a thriving economic core. Near the entrance to the massive Portuguese Fort Jesus, where the city wall would once have stood, air-conditioned buses today discharge tourists, a few of whom venture down the well-trodden paths of Old Town. Here the streets appear chaotic, and each turn brings unexpected surprises to the outsider **(Figure 5.3)**. At times the

Figure 5.3. A street scene in Old Town Mombasa (Mbabak Hinawy Road), looking toward the old Mandhry Mosque. (Courtesy of Thomas Gensheimer.)

roads are cavernous passages, framed by wooden balconied buildings which press even further on the narrow roadway. In other places, they open up to reveal informal squares or the occasional whitewashed mosque and minaret. The general impression of this older area of the city, compared with the constant din of cars and people in the modern part, is that it has been deserted, replaced and rejected.

This image of the modern city highlights the present predicament of Kenyan identity. Kenya is an Indian Ocean power (through Mombasa) which has been behaving as if it were an Atlantic power (through Nairobi). Despite Mombasa's claims to be a modern global city, its residents believe it has been neglected by the country's leaders in favor of Nairobi. As a result, its concrete buildings appear weathered from the tropical climate, automobile exhaust, and the grit that blows through the streets on the ocean breeze.

Perhaps of all African countries in the postcolonial era, Kenya has evolved the most wide-ranging relationships with the leading countries of Western Europe and North America. For a while Kenya was the darling of the West, not only because of its tourist attractions but also because of its strategic importance. Ironically, Kenya's strategic value to the West lay precisely in the fact that it was an Indian Ocean power. It was located along one of the great sea routes of the world in close proximity to the oil reserves of the Middle East. As such, it became critical as a potential staging place for the rapid deployment of Western forces. It was vital that Kenya remained pro-West.

A paradox emerged. Even though the West flirted with Kenya because of its location on the Indian Ocean, Western flirtations turned Kenya's eyes away from the Indian Ocean. The country became excessively Western oriented and inadequately attentive to its Asian neighbors. The West embraced Kenya partly because the West valued the Indian Ocean; but Kenya embraced the West and turned its back on the Indian Ocean. Mombasa was part-and-parcel of the paradox.

Now the honeymoon is over. Kenya's intimate embrace of the Western world at the expense of all other relationships has at last to be reviewed. Kenya is a country which in the past has been lukewarm toward pan-Africanism, hostile to the People's Republic of China, profoundly distrustful of the Arabs, indifferent toward India, and cool toward the Soviet Union (or Russia). For the bulk of the first 25 years of the Kenya African National Union (KANU), Kenya's most intimate friends have been the Western powers (and Israel). Kenya has often been amply rewarded by these particular friends.

Now that the Cold War is over, the West has itself begun to review its relationship with Kenya. Many Western countries have decided that they no longer have to humor the present KANU government simply because it is anti-communist. Western governments have at long last become friends of the cause of democracy in Kenya. Their pressures helped force President Daniel Arap Moi to accept the principle of multipartyism — though one must also not underestimate the impact of domestic democratic pressures on President Moi's change of mind.

Whether KANU or some other party comes to power after the multiparty election, it is time that Kenya, too, embark on a fundamental review of its wider global options. Just as Kenya is now of lesser strategic value to the West with the end of the Cold War, so should the West become less fundamental to Kenya's global design in the wake of those wider developments.

On its ocean front, in Mombasa, Kenya today faces the Asian rather than the European world. Until the 1990s Kenya has been far more sensitive to the imperatives of history (the history of being an ex-colony of Britain) than to the imperatives of geography (proximity to Asia and the Middle East). It is time to redress this imbalance between postcolonial historical continuities and the older dictates of geographic proximity, but first Kenya has to recognize that it is indeed an Indian Ocean power.

Indian immigrants came to colonial Kenya as long ago as the building of the railway line to Uganda at the beginning of the century. In the past Kenya has tended to regard this South Asian population as a politico-historical liability. Kenya should now learn to recognize its Asian population as an geo-economic asset. In population, India is now the second largest country in the world after

China — and therefore destined to become a superpower in the future. Likewise, Pakistan is one of the world's leading Muslim and Third World powers. Can Kenya work out a special relationship with both India and Pakistan? Kenya's Asian population can help forge economic, technological and educational ties.

The Kenya coast is so culturally Indianized that the food culture of the Waswahili uses a lot of Indian spices, the music of the Coast Province shows evidence of South Asian impact, and Indian films are heavily patronized. This cultural influence of South Asia in Kenya could be converted into diplomatic currency in Kenya's economic relations with India, Pakistan and Bangladesh. Trade between Kenya and those countries could also be enhanced if careful thought was given to the nature of economic exchange in the future. Kenyans could learn a lot about intermediate technologies from their South Asian neighbors.

KENYA AND THE ARABS

The KANU government's history of anti-Arabism has been even more costly to the country. Kenya's pro-Zionist orientation was often excessive and unnecessarily extreme. Arab travelers were often subject to humiliating visa and customs policies. And yet, arguably, Kenya could have been the most attractive country for Arab investment in the whole of Black Africa. It is geographically close to the Arab world; it shares borders with two members of the Arab league (Sudan and Somalia); and it is the most stable country in Eastern Africa. Culturally, Kenya has an Arabized population among its citizens (the Waswahili, the Somali, and the other Muslims groups). Economically, it has been prosperous since independence and has exported both skilled and semi-skilled workers to the Arab world (including a former Kenyan ambassador to Zaire, who later became the ambassador of the Sultanate of Oman to the People's Republic of China).

Kenya could easily have become a major magnet of Arab money had the KANU government not been so disproportionately pro-Zionist. Fences have been partly mended since the mid-1980s, but Kenya-Arab relations could do with a lot of improvement. Kenya's pro-Zionism was once so extreme that Kenya became the only Black African country that was a target for so-called "Arab terrorism." The Norfolk Hotel in Nairobi was bombed by an outraged Arab when Kenya was perceived as a military ally of Israel.

Future Kenyan governments will have to review their relations with the Arab world as a whole. In terms of population, the bulk of the Arab world is within Africa. The most populous Arab country is Egypt, a fellow member of the Organization of African Unity (OAU). The largest country in Africa in square miles is Arabized Sudan, with which Kenya shares a border. In addition, Kenya's armed struggle for independence under the Mau Mau nationalists overlapped in time with Algeria's armed struggle for independence under the National Liberation Front. Algeria is an important member of both the OAU and the League of Arab States.

Just as the concept of Eurafrica makes sense in Europe's special relationship with African states, so should the concept of Afrabia make sense in the context of the history and geography of African-Arab relations. Kenya has been a key country in the partnership of Eurafrica. Kenya could also be a key country in the solidarity of Afrabia. Kenyan goods, Arab oil, and petro-dollars could do business with each other under more enlightened leadership from Nairobi.

KENYA, CHINA AND JAPAN

China probably invented tea. The Swahili word *chai* is in all likelihood Chinese in ancestry. Will Kenyans be selling more tea to China in the years ahead? This may sound like taking coal to Newcastle, or palm dates to Muscat in Oman, but the issue simply dramatizes the immense potentialities of the Chinese market for Kenyan products. The Chinese have begun to diversify their tastes and consumption patterns — even to the extent of experimenting with new varieties of tea (Indian tea as well as Chinese). Some meat products and canned tropical fruit may also have a significant potential market in the People's Republic of China. On a visit to Beijing in the 1980s, I was astonished by the number of businessmen and advertisers I met.

For at least two decades, the KANU government has treated the People's Republic of China with greater suspicion than that shown by the majority of Western states. The Kenyan government has often been more anti-communist than Western countries, and Kenyan courts still tend to regard works by Mao Tse-tung as "subversive literature." In a back street of Nairobi, I once saw an exceptionally beautiful and elegantly bound edition of Mao's works in a dustbin. As a scholar, I was strongly tempted to rescue it from the trash bin and take it to a library where it really belonged. But as a citizen, I was reluctant to break the law — even though I regarded the law as both unjust and absurd. In the end, the citizen in me triumphed over the scholar. With great reluctance on that occasion I let Mao's works rot among the garbage in the back streets of Nairobi — and walked away in painful ambivalence.

Even in China, only a minority may still believe in Maoism — although many more continue to admire the man's contribution to the history of China and the world. A future Kenyan government has to recognize that one need not ban a book simply because one disagrees with it. Future Kenyan governments may also have to be more diplomatic and considerate in their treatment of the People's Republic of China.

Kenyan businessmen are already eager to do business with Japan if at all possible. The next Kenyan government should go further than that. It should encourage some Kenyans to study Japan as a model of modernization and development. Here is one non-Western country that has taken on the West at its own technological game — and given the West a run for its money. What is there in Japanese culture that permitted custom and tradition to be major engines of modernization? The next Kenyan government should encourage research into "the culture of modernization and development." If the Japanese could creatively and selectively learn from the Western world, can Africa selectively and creatively learn from the Japanese? The Koreans have learned from the Japanese. Can Kenya become Africa's South Korea?

BETWEEN GLOBALISM AND PAN-AFRICANISM

As for Kenya's relation with the rest of Africa, a particularly strange dialectic has been at work. In many matters Kenya has been more globalist than pan-African. It has hosted such major global assemblies as that of women in 1985, and it has been the site of major international developmental conferences sponsored by the World Bank and the International Monetary Fund. Nairobi has not been in the forefront as a host to the Organization of African Unity. Nairobi is the global capital of the United Nations Environmental Program. It is also the African capital of the World Council of Churches through its affiliate, the All-Africa Conference of Churches. But Nairobi has hesitated to serve as a capital for purely African organizations or activities. The headquarters of the Organization of African Unity are in Addis Ababa Ethiopia; those of the African Development Bank are in Abidjan, Ivory Coast. The headquarters of the Association of African Universities are in Accra, Ghana; and the headquarters of the Africa Leadership Forum are in Ota, Nigeria. Nairobi is the most accessible capital in Black Africa, but it has not tried very hard to serve as a center for major pan-African activities. In the first 29 years of its postcolonial life, Kenya has at times tried to be globalist; but it has not tried hard enough to be pan-African.

Kenya's virtue of globalism has been flawed by its excessive preoccupation with the Western world. Fortunately, its neglect of pan-Africanism has been somewhat ameliorated by the respect in which its two presidents have been held as elder statesmen on a continent that is culturally sensitive to the elder tradition (*heshima ya uzee*). Both the late Mzee Jomo Kenyatta and President Daniel Arap Moi established records of earnest mediation in inter-African disputes: Zaire in the 1960s, Angola in the 1970s, Uganda in the 1980s, Mozambique and Somalia in the 1990s. The efforts by Mzee Kenyatta and President Moi to reconcile warring Africans have seldom been successful, but it is to the credit of these two presidents that they have been prepared to try. Kenya's pan-African balance sheet may be in the red, but the roles of these important men as mediators are on the credit side of the country's pan-African account.

As for the sad disintegration of the East African Community, there is enough blame to go around among the three old members: Tanzania, Uganda and Kenya. Charles Njonjo is widely regarded as the person responsible for the final act of destruction, the *coup de grâce*. It is true that Mr. Njonjo was no friend of the concept of an East African Community, but by the mid-1970s the community had been greatly undermined and damaged by the policies of Tanzania and Uganda, as well as Kenya. The region as a whole remains guilty of having destroyed what stood a chance of becoming the most successful case of regional integration in the entire world. Can the East African community be revived before the end of the twentieth century?

CONCLUSION

As Kenya tries to restore balance in its pan-African policies, and as it attempts to reestablish its credentials as an Indian Ocean power through Mombasa, what is to happen to its special relationship with the North Atlantic? Kenya still has many friends in the Western world. The number of supporters it may have lost by becoming less strategically valuable to the West may be compensated by the new friends Kenya will make as it becomes truly democratic. In December 1991 I was privileged to meet the Prime Minister of Norway, Ms. Gro Harlem Brundtland. Although diplomatic relations between Norway and Kenya were indeed broken, she emphasized to me that it was Nairobi that had decided on the diplomatic break. "Norway believes in a democratic Kenya, not in isolating Kenya." President Daniel arap Moi later capitulated to international and domestic pressures — and accepted the principle of multiparty politics. Although Kenya may lose strategically conscious Westerners among its old partners, it may gain democratically conscious Westerners among its new friends.

Kenya may consolidate its role as an Indian Ocean power through Mombasa — without losing all the Western gains it made when it behaved like an Atlantic power through Nairobi. A new equilibrium can indeed be achieved but this will require new political will, a greater sense of history, and a deeper sensibility to geography.

Much older than Kenya as a country, of course, is Mombasa as a city. Kenya is a twentieth-century political entity; Mombasa is a phenomenon of at least a millennium. Just as Mombasa is an ancient city and Kenya is a young nation, Mombasa is a global city while Kenya is still grappling with the basic dilemma of its geographic contiguity with the Indian Ocean and its historical continuity with its Atlantic legacy.

NOTES

An earlier version of this chapter was published as "Mombasa: Three Stages towards Globalization," in Anthony King (ed.), *Re-Presenting the City: Ethnicity, Capital and Culture in the Twenty-First Century Metropolis*, New York, NYU Press, 1996. The author retains copyright over the earlier version.

1. See N. Chittick and R.I. Rotberg (eds.), *East Africa and the Orient*, New York, Africana Press, 1975. Also see B.A. Datoa, "Influence of Monsoons on Movements of Dhows along

the East African Coast," *East African Geographical Review*, vol.12, 1974, pp.23–33.

2. See, for example, C. Sassoon, *Chinese Porcelain in Fort Jesus*, Mombasa, National Museum of Kenya, 1975.

3. See R. Skene, "Arab and Swahili Dances and Ceremonies," *Journal of the Royal Anthropological Institute* (London), vol.47, 1917, pp.413–434.

4. See T.O. Ranger, *Dance and Society in Eastern Africa, 1890–1970: The Beni Ngoma*, London, Heinemann, 1975.

5. See, for example, J. de V. Allen, "Ngoma: Music and Dance," in Mtoro bin Mwinyi Bakara and J.W.T. Allen (eds.), *The Customs of the Swahili People: The Desturi za Waswahili*, Berkeley, University of California Press, 1981, pp.233–246.

6. See, for example, R.L. Pouwels, *Horn and Crescent: Cultural Change and Traditional Islam on the East African Coast, 800–1900*, Cambridge, Cambridge University Press, 1987.

7. See J. Strandes, *The Portuguese Period in East Africa*, Nairobi, East African Literature Bureau, 1961.

8. See also G.A. Akinola, "The Mazrui of Mombasa," *Tarikh*, vol.2, no.2, 1968, pp.26–40; and C.R. Boxer and C. de Azevedo, *Fort Jesus and the Portuguese in Mombasa 1593–1929*, London, Hollis and Carter, 1960.

9. J. Plamenatz, *On Alien Rule and Self-Government*, London, Longmans, 1960, p.111.

6

Rethinking Heritage Politics in a Global Context: A View from Istanbul

Ayfer Bartu

On August 25, 1994, I participated in a tour of Beyoglu/Pera (the first "Europeanized" quarter of Istanbul) organized by the Turkish host committee of the AESOP (Association of European Schools of Planning) Congress, which was at the time holding a meeting in Istanbul on "Planning for a Broader Europe."[1] The tour was one of five in the city organized by the committee, and undoubtedly the most popular. It was advertised as "A Walking Tour of Beyoglu/Pera: A Vanishing Grace." According to the tour pamphlet, it showcased a special era in the city's history:

> The history of Istanbul is generally divided into three major parts: The Ottoman period follows the Byzantine and leads up to that of the Republic proclaimed in 1923. Yet there exists also a secondary period which has been called, rightly or wrongly, *la belle epoque*, and which deserves to be recalled on its own right. This is the time when, for a fleeting historical moment, Istanbul missed becoming a European metropolis through its cosmopolitan and multicolored aspect which earned it the name *Darsaadet*: the Port of Felicity.[2]

The tour was given by a prominent historian, and there were around thirty participants, many of whom were European (Norwegian, Dutch, British, French, Austrian, Italian) urban planners. As we walked down the main street of the quarter (İstiklâl Street), our attention was drawn to churches, consulates, and Russian restaurants, and we took several breaks at cafes which had been renovated and opened in the mid-1980s **(Figure 6.1)**. At the end of the tour, we walked down Galata Tower, a structure built by the Genoese, toward the Golden Horn. As we stood along this famous waterway, to our right was Istanbul's historical peninsula with Topkapı Palace and several Ottoman mosques dominating the skyline, and to our left stood Galata Tower and a Catholic Church. Pointing to Istanbul's historical peninsula, a British planner told a Norwegian one: "You see that mon-

Figure 6.1. A view from "Grande Rue de Pera" in the late nineteenth century. (Postcard.)

umental architecture; that is real Istanbul — the authentic one." Then, pointing to the hills above Galata Tower, he continued: "This is fake Europe, an imitation." A French planner joined the conversation. "You have a point. But I feel safer here. It feels more familiar. This is more like nineteenth century Europe. Maybe one should stay here, and visit those monumental sites as a tourist."[3]

This sense of ambivalence toward what Beyoglu represents is shared by Istanbulians. Is Beyoglu part of Istanbul's heritage? If so, is it "Turkish," "European," or "world" heritage? Does it represent a "vanishing grace," as the tour organizers claimed? And with its nineteenth-century "European"-style buildings, does it really symbolize "the time when Istanbul missed becoming a European metropolis"?

Emerging research on the politics of heritage has drawn attention to the ways history, memory, and the past may be contested and negotiated within the politics of the present.[4] Heritage, it is now understood, may acquire complex and multilayered meanings across different sites and periods. In the pages that follow, I examine how the cultural heritage of Istanbul has been contested in recent debates over the preservation and revitalization of the Pera/Beyoglu district. My focus is on two periods — the 1980s and 1990s. During the 1980s, the meaning of Beyoglu's heritage became central to the struggle over its revitalization, undertaken as part of a larger, government-sponsored effort to transform Istanbul into a global, world-class city. Then, in the 1990s, the district's heritage once again became a site of struggle, as various groups sought to respond to the Islamist Refah Party's victory in the 1994 municipal elections. In relation to these events, I attempt to show how Beyoglu/Pera's past has been used as a political symbol by different groups in their attempts to sway both local and global opinion; I discuss the ways in which different readings of the past inform and shape present debates over cultural heritage; and I try to draw attention to the centrality of heritage politics in the globalization process.

This essay is organized into two main parts. The first looks at views of Beyoglu/Pera before the election of 1994, and attempts to describe the complexity of meanings the district held for different groups, particularly with regard to the revitalization project of the 1980s. The second part examines the scramble to adjust these views in light of the victory of the Islamist Refah Party in the 1994 elections, a major historical fault line in attempts to define the urban space of Istanbul.

BEYOGLU/PERA

> Pera and Galata, the only busy quarters of Constantinople, are of no land and of all lands, each country administrating its own laws, exercising its own religion, circulating its own money, distributing its own letters. Here are the various banks, consulates and embassies, bazaars, churches and chapels, including the dancing dervishes, who display themselves on stated days. Water sellers, porters, sedan-bearers, who carry you out to dine in a delightful sedan-chair, like the old ladies at Bath; Albanians, wood-cutters from Asia Minor, Persian donkey-drivers, Croats and Native Turks, form a polyglot population unparalleled in the world. If I were to enumerate all the articles sold in the street, or all the languages spoken, I could fill pages which would interest no one.[5]

Beyoglu/Pera has always been a marked neighborhood within Istanbul. Established as a Genoese trading colony in the thirteenth century, it had autonomy from the Byzantine Empire and controlled trade routes in its immediate vicinity. Byzantines referred to the area as Pera, which means "beyond" or "far away" in Greek.[6] It was "far away" for the Byzantines in several senses: geographically, it was located on the other side of the Golden Horn; and because of its autonomy, the Byzantines did not have control over it.

When the Ottomans conquered Byzantine Constantinople in the fifteenth century, this Genoese colony formed an alliance with the new rulers and managed to maintain its autonomy as a trading colony. Then, at the beginning of the nineteenth century, the native minorities of the Ottoman Empire — Greeks, Armenians and Jews — started to move into its neighborhoods. Eventually, it became that part of the city where much of the non-Muslim population lived, along with people who were referred to as "Levantines" (the descendants of Europeans settled in Istanbul, some of whom were the offspring of intermarriages with other minorities).

For the Ottomans, especially in the nineteenth century, Pera represented "Frank" Istanbul,[7] and they referred to it as "Beyoglu." However, by the nineteenth century it had also become the financial and entertainment center of the city, the area where embassies were located, and where the Pera Palace Hotel was built for passengers on the Orient Express. By this time Pera was also Istanbul's most affluent neighborhood, home to most of the empire's non-Muslim merchant families **(Figure 6.2)**.

Later, toward the end of the nineteenth century, in the course of Ottoman efforts to transform Istanbul into a Western city, the district was designated as the location for experimental reforms aimed at creating a new urban planning model.[8] The selection was not accidental, given Pera's popularity and its largely European and non-Muslim population. Thus, Pera became Istanbul's showcase Europeanized quarter, dominated by symbols of modern living such as office buildings, banks, theaters, hotels, department stores, and multistory apartment buildings.

With the transformation of the Ottoman Empire into the Turkish nation-state after World War I, Istanbul lost its status as capital. In contrast to the new capital of Ankara (symbol of the new Republic, of secularism, and of enlightenment), Istanbul symbolized the decadence and corruption of the Ottoman Empire and its entrenchment in Islam. Yet even during this period, Pera remained a symbol of civilization for the Turkish bourgeoisie — Europe in Istanbul. However, over the years that followed, given the changes in its population as non-Muslim minorities left the city and the country, it also came to serve as a symbol of the transformation from empire to nation-state.[9]

Then, starting in the 1950s, most of Beyoglu's buildings came to be occupied by squatters who migrated from the countryside. With its nightclubs and bars, it also began to cater to nouveau-riche merchants from outside the city. Yet, even though the function and population of the district had changed drastically, it still contained the most well-preserved nineteenth century urban fabric in the city.

Remaking of Pera in the 1980s

A *Financial Times* writer had the following observations on his visit to Istanbul in the early 1990s:

> The new wealth and confidence of Istanbul are being translated into art and culture. There is a new passion for Islamic and Ottoman art. Istiklal Street is again becoming as elegant as it was in the nineteenth century, when it was the cosmopolitan, French speaking Grande Rue de Pera.[10]

As Kevin Robins and Asu Aksoy have suggested, what this writer observed was what he was intended to observe: an "elegant," "cosmopolitan," sanitized image of a city "appropriate for international acceptance and recognition."[11] Such a transformation was partly due to the efforts of the government of the

Figure 6.2. A view from İstiklâl Street (Independence Street) in the late 1990s. (Photo by author.)

center-right Turgut Özal, who undertook the challenge of transforming Istanbul into a "global," "world-class" city during the early 1980s. Under his government, Turkey opened its gates to global commodity capitalism, lifting tariffs on foreign goods and allowing more multinational companies to do business there. The city of Istanbul was central to this intensified activity, eventually emerging as a showcase of transformation with shopping malls, five-star hotels, high-rise buildings, and entertainment centers. As David Harvey has suggested:

> . . . under the speeded up circulation conditions of hypermodernity, where little escape commodification, there is nothing so big, so encompassing, so overshadowing, so enormous in its implications, that it cannot be commodified, that it cannot be represented as a consumption spectacle. Through imagery and massive architectural projects, even entire cities or metropolitan areas may be served up as market objects and all-encompassing spectacles.[12]

In their efforts to transform Istanbul into a global city, the government of Özal and his Istanbul mayor, Dalan, became keen on "transforming Istanbul from a tired city whose glory resides in past history, into a metropolis full of promise for the 2lst century."[13] Accordingly, its image as "the East in the West, and the West in the East," and the "gateway to the Orient" was mobilized to market it to mainly foreign investors and tourists. Accompanying the marketing campaign was a series of highly publicized urban-renewal projects through which Istanbul was to become a consumption spectacle. Beyoglu, which had been the site of urban reform in the nineteenth century, was included in these new urban-renewal projects through a massive revitalization plan. The main idea was to turn İstiklâl Street (the main street of the district) into a pedestrian artery, and to open a new street (Tarlabaşı Boulevard) parallel to it. The new boulevard would connect the central Taksim Square (Beyoglu's main square) with the Atatürk Bridge across the Golden Horn. But the opening up of such a new boulevard would require the demolition of many nineteenth-century buildings and the relocation of thousands of people.

As was the case with most of Dalan's projects, the Beyoglu revitalization/demolition project was controversial, and what were referred to as the "Tarlabaşı demolitions" dominated public discourse throughout the 1980s. The most vocal protagonists in this debate were Dalan on one side, and the Chamber of Architects (MMOB) — the main organized opposition to Dalan's projects all over Istanbul — on the other. The struggle between the two camps was typically framed as between "modernizers" (referring to The National Party (ANAP) and its right-wing rhetoric), and "conservationists" (who were associated with MMOB and the left-wing). Beyond such a dichotomous conceptualization lay a number of much more complex and multilayered differences. At one level these demonstrated the many ways that history and its relics (and different readings of them) may be used as symbols in contemporary political struggles. As with every revitalization attempt, this one provoked different readings and interpretations of the past, as different actors and groups advocated different narratives about Beyoglu's physical landscape, which could also be read as commentaries on its present social, political and cultural meaning. Thus, questions such as how Beyoglu used to appear, which Beyoglu should be revitalized and for whom, and what Beyoglu represented and to whom became critical political issues. Meanwhile, and at another level, the struggle over the meaning of Beyoglu brought home general problems about Istanbul's urban culture, urbanity, and urban identity. Within contemporary power relationships, what did it mean to be "Turkish," "European," or "Istanbulian"?

On the surface, positions over Beyoglu were fairly clearly drawn. For Mayor Dalan, Beyoglu was a place that needed to be cleaned, rehabilitated, and — in parts — demolished. Toward these ends, the proposed highway project would take care of traffic congestion in the inner city and clear Beyoglu of illegal prostitution and drug trafficking. Any canon of historical preservation was seen as an impediment to such a drive to develop and transform Istanbul into a world city. According to Mayor Dalan:

> We are against the preservation of historical sites if they impede development. We are ready to undertake any punishment in order to serve this city. We will implement the plan and if needed, we are ready to undertake any penalty. There are no historical buildings in Tarlabaşı, at least the ones we have demolished are not historical. We will continue the demolition.[14]

For the MMOB, Beyoglu, and specifically the Tarlabaşı demolitions, became both the physical and symbolic site of a political struggle against the policies of the center-right party, the ANAP. Yet, although the main focus of MMOB's campaign was the illegal nature of the demolitions and concern about land speculation, positions within the group diverged over more symbolic issues. Thus, according to the head of MMOB:

> What they are doing is selling the city to multinational corporations. These places will be sold to private speculators and some will turn into multistory parking lots which in turn will draw more traffic into the area rather than solving the traffic problem.[15]

Others were more concerned about the racist overtones of the demolition rhetoric:

> Besides the fact that these people don't know anything about urban planning, they are also racists. They demolished these buildings claiming that they have been the houses of Armenians and Greeks. One of them gave a speech next to a bulldozer covered with a Turkish flag. And this kind of thing does appeal to people; remember what happened on September 6th and 7th.[16]

Demolition was also critiqued on the basis of the buildings' architectural value. "This architecture is a synthesis," one architect argued. "It is neither European nor Ottoman; it is Levantine architecture."[17] Opponents of demolition eventually proposed an alternative project to save the buildings. Besides preserving architectural value, they claimed the architectural "synthesis" the buildings represented could be used by the city administration to market the city to tourists. According to one opponent of the demolitions:

> It is a disaster to pass a transit highway through the middle of the city. This is the only area in the city where you can find well-preserved buildings from the same period. Istanbul is a city within Europe with Eastern culture, we have to use this potential.[18]

Given such internal divergences, it was remarkable that MMOB managed to present a unified front. But the debate became even more complicated when one considered the splits among other left-wing groups. For example, among some left-wing critics, rather than being a "synthesis," the architectural heritage of Beyoglu symbolized nineteenth century European capitalism. According to one prominent left-wing novelist:

> There are people who think that old Istanbul is destroyed. The thing which is destroyed is Pera, and Pera has nothing to do with Turkishness. I don't see any problem with demolishing these buildings. Moreover, all of them were designed by Armenian architects. They are built in Western style. All are imitations. Their demolition will not be a great loss. They are the product of slavery years. Just as in India today British buildings are irrelevant, it is really unnecessary to preserve these buildings where the managers of the Ottoman Bank, Jewish families, and Turkish representatives of foreign investors used to live, and which are now being used for illegal prostitution. In other words, these places, which were all inhabited by foreigners, reminding Europe in the Ottoman Empire, are not the product of Turkish culture.[19]

This account is striking in its apparent contradictions: a left-wing position in agreement with right-wing rhetoric, disinheriting the heritage of the area, claiming it had nothing to do with "Turkishness." Such a view demonstrated both the power of the Turkish nationalist project and its ambivalent relationship with Europe, and it reflected the historical reality that although the Ottoman Empire was never colonized directly, it did become part of growing capitalist networks, especially in the nineteenth century. As Ayse Keyder and Caglar Öncü have argued, the empire was integrated with European markets through the intermediary positions of its indigenous (local) non-Muslim communities.[20] In Istanbul the penetration of European influence was symbolized by the districts of Galata and Beyoglu, the financial center and the residential area, respectively, for groups involved in these commercial activities.[21] Although the Turkish nationalist project was willing to identify with European civilization, it was also defined as a struggle against European imperialism. The local non-Muslim minorities eventually became the symbol of this imperialism and the targets of "ethnic cleansing."

If one turns to another set of narratives, also very much a part of the public debate over Beyoglu, still further versions of what the district symbolized become evident. According to these competing views, Beyoglu was at once a symbol of "civilization" and "elegance"; a "brothel"; and a "foreign" heritage appropriated for "national" reasons.

For people who considered themselves real Istanbulians, the decline of Beyoglu, like that of many of Istanbul's other neighborhoods, represented the "ruralization of the city."[22] Although the Turkish nationalist project was, and continues to be, based on the glorification of the peasant, for urbanites the value of peasant culture took a different meaning when the peasants came to the city to stay. According to the chronology offered by these urbanites, the change in Istanbul started in the 1950s with the beginning of mass migration from the countryside, and the city was subsequently "conquered" by immigrants from Anatolia. One older man from a well-known Istanbul family described the change as follows:

> Istanbul was conquered again in the 1950s, 500 years after Sultan Mehmet's victory, by the Anatolian invasion. These people brought their own civilization to my city, instead of trying to adapt to ours. I am sure that

> none of these people have ever been to an exhibition in their lives; all they think about is getting enough money for a summer house. We became a nation of *lahmacun* eaters. Fifty years ago no one in Istanbul knew what *lahmacun* was — or if we did, we called it pizza.[23]

Such a view expressed nostalgia for a "lost city," symbolized by the romantic view of the Grande Rue de Pera as a place where one had formerly seen "only well-dressed and well-behaved ladies and gentlemen."[24] Such a sanitized and nostalgic version of Beyoglu's past was aligned with the abundance of best-selling books and other nostalgic literature on Beyoglu at the time. The authors of these books and their audience would have agreed that something had to be done to revitalize the quarter to re-create its former elegance. And they did not object to demolitions, if they would help restore some sense of "old Beyoglu." With great contempt, one writer who used to live in the area said:

> Beyoglu used to smell of perfume and sesame, now it smells of *lahmacun*. When I was a child, "to go to Beyoglu" was something. We would wear clean and neat clothes, and shave. Beyoglu was a mark of civilization. If these demolitions and closing down İstiklâl Street to traffic will help to return the old ambiance, I don't have any objections to them.[25]

On the other hand, many had reservations about the particulars of the revitalization strategy for what might be called sociological reasons. What had once made Beyoglu unique was its people. Preserving or restoring the buildings alone would prove futile, because the buildings were no longer populated by the same people. Thus, the spirit of the place would be lacking, even after the area was preserved. Ultimately disappointed with the outcome of the revitalization project, one old inhabitant complained:

> These were places like the cafes in Paris, where people would sit and talk with friends, hold meetings, do business and make appointments. These new cafes in Beyoglu are trying to recreate this era, but they are just not the same. The spirit is lacking. I would like to live back in the golden days of Pera at the turn of the century — the days when Jewish, French, Greek and Armenian people all lived together.[26]

However, another popular view of Beyoglu at the time was that it had turned into a "brothel." And some were puzzled by the rhetoric of elegance that was often used to describe the old quarter. For example, an actor who had lived in Beyoglu for 23 years, said:

> Beyoglu has always been one of the most popular red-light districts of the city, and still is. I don't really understand what elegance these people are talking about, and I don't see any reason why this neighborhood should not be cleaned up. I don't have any objections to the demolitions.[27]

A compromise position was offered by an activist who was highly critical of Dalan's projects.

> We need neither the nostalgic nor the brothel version of Beyoglu. We need a national and clean Beyoglu. One cannot argue that since people are gone, we should also get rid of the buildings. We are neither Franks nor Levantines. We have to restore these buildings and open them up for tourists. We should be the hosts and Levantines should be the guests.[28]

These different positions reveal just how complicated the various meanings people attributed to Beyoglu were. Such complexity provoked a number of questions. What was Beyoglu all about? What did its built environment represent? Could one think of it as "heritage"? If so, was it worth preserving? Who was the audience for its preservation and revitalization? Some argued the quarter should be demolished because of what it represented. Others suggested it should be demolished to re-create what it had been (although many had reservations about what this might entail). Still others argued that it should be preserved because of what it was, emphasizing its architectural value and potential for tourism.

For their part, business owners in the area were highly selective in their use of these multiple interpretations and cultural assertions. Yet Beyoglu's history, historical value, and symbolism turned out to be important resources for them, too. And they generally split on the issue in line with their interests: some accepted Mayor Dalan's offer for new shops in other areas, and others formed alliances with MMOB to resist the demolitions. A business owner who had one shop on İstiklâl Street and another on Tarlabaşı Boulevard (in the only protruding building saved from the 1980s demolitions) recounted his experiences:

> At that time, I had great support from MMOB. They have been telling me that I was the symbol of their resistance for their struggle against Dalan. But it wasn't that all the shop owners were unhappy about the demolitions. Dalan had tempting offers for them and they accepted them. Maybe they were right after all; now I regret it. . . . Look, I have all the documents from Ministry of Culture showing that my building is a historical one, but what does Dalan care about history? I was able to save my building with this document, but I still have to deal with all the legal cases. Now, what do I have after all those years of struggle? I had a heart attack, developed asthma. . . . I don't have MMOB anymore to support me, their problem was with Dalan. After Dalan lost the elections, MMOB was gone too. I still have to pay all the legal expenses for the pending cases.[29]

It should be pointed out that the historical value of this business owner's building, which he documented in order to save it from demolition, had since become a burden for him. Once Dalan lost the local elections, the owner found himself alone in dealing with the ongoing legal cases. And because most of the other business owners had accepted the offer for new shops elsewhere, he had not been able to conduct business, either. He was left with only a historical building with no economic value.

The highly selective nature of remembering the multiple histories of this district may best have been captured by a Levantine who had been living in Beyoglu for thirty years:

> The question is, within all this hysteria of nostalgia over Beyoglu, what are we really yearning for? Which Beyoglu? Beyoglu itself has changed a lot. If we want Istiklal Street of the 1900s, it was a muddy and extremely narrow street. When we read the travel accounts of European travelers who visited Istanbul in the second half of the last century or the beginning of this century, we see that Beyoglu was not really a clean place. It was even worse than today. There is the period of Capitulations: do we want that Beyoglu? There is the Beyoglu of the times when the "White Russians" rushed to Beyoglu: that's a different one. There is the Beyoglu of the war period and the period of September 6th and 7th. Or do we want all? We can't really decide.[30]

A Different Chronology: The Legacy of Istanbul and Pera in the Islamist Imagination

A completely different reading emerges if one examines the meaning of Beyoglu within Islamist discourse. As a movement that challenges the Turkish nationalist project (which defines itself in opposition to the Ottoman past), the Islamist movement has attempted to revitalize and resurrect the past. In its reading of history, everything Ottoman serves as a symbol of Islamic ideology and can be used politically to challenge the Turkish secular enterprise. If the Islamist aim is to resurrect a lost, glorious Ottoman past, then Istanbul, the glorious capital of the former empire, was necessary as a key symbol.

The Islamist discourse defines a chronology of changes in the city that is quite different from the one proposed by secularists. For Islamists, the real history of Istanbul started with the conquest of Constantinople in 1453 by Sultan Mehmet. In this historical narrative, Istanbul came to represent the organic unity and justice of Ottoman (Islamic) rule. This view was well expressed in a recent photograph album, *Istanbul: A Glimpse into the Past*, published by the Research Center for Islamic History, Art and Culture.

> These are the images of a period in which this city of the secret tales and unknown beauties was just beginning to be Westernized. The city of Istanbul, known as "Dar Saadet" (the door to happiness) in those days, was truly *a realm of happiness*, where the aesthetic taste of life which matured with time, made its inhabitants happy. . . . In this city, where each neighborhood had a different beauty, atmosphere, and view, buildings and people made up *a magical unity*

> *within the simplicity of daily life.* The disturbing pressure of a geometric order made up of grand boulevards, large parks, monumental fountains, palaces, memorials and stone bridges laid out in the logic of urbanism, and shared by so many famous contemporary cities was absent in Istanbul. . . . You will see in these photographs, Istanbul in its *authentic beauty* which survived until the beginning of the process of Westernization [emphasis mine].[31]

The pristine, authentic city depicted here, characterized by magical unity and harmonious existence, embodies all the characteristics of the "Islamic city." Thus, Ottoman history was Islamic history, and Istanbul became the "Islamic city." Such authenticity was lost with the Westernizing reforms of the nineteenth century. As one Islamist intellectual put it:

> There are three Istanbuls. One is the Byzantine Constantinople, the second is the Ottoman-Islamic Istanbul, and the third one is old Pera. This last was never Islamicized and its extension now is quarters like Şişli, Mecidiyeköy, Levent, Bebek, which are inhabited by modern business offices, skyscrapers, entertainment centers, big hotels, all buildings symbolizing the modern world. *The essence of Istanbul is the part which reflects its Ottoman-Islamic identity* [emphasis mine].[32]

Given that Pera was the symbol of order and modernity for nineteenth century Ottoman reformers, it is not surprising that, within the Islamist discourse, this is exactly the place where the city's "urban problems" began. As another Islamist intellectual suggested: "Pera, which is marketed as the symbol of Istanbul's civilization and elegance with words like 'culture,' 'civilizational heritage' and 'nostalgia,' is actually the place where our contemporary urban problems emerged."[33] In other words, Beyoglu represented nothing less than cosmopolitan degeneration and foreign cultural invasion. "Istanbul has been owned by the Turks since 1453 with one exception, Pera, which has always been a sore in the Empire's brain, a Frank sore."[34]

The theme of ownership was also addressed by another Islamist intellectual in discussing the urban transformation of Istanbul. In the following excerpt from a TV interview, he reflected on the new identity of the city:

> Starting with the 60s, there is an incredible immigration. It is a problem one cannot deny. One-sixth of Turkey lives in this city. But I also think that there are certain advantages to this. This development provides a new color to city life. Most importantly, these people besiege the city with mosques. Some don't like their architecture, but that's another matter. Once they tried to transform this city into a Turkish one. Now you don't need to make an intentional effort for this, it is already there. It is distinct now. There is a new lifestyle now. Istanbul used to absorb this migration. It is a city of migrants anyway. But it used to be a melting pot. The dominant culture absorbed the

> rest. There was something called the Istanbulian way and everyone adapted to it. But now, Istanbul has a culture which has to be redefined and reformulated. It tries to redefine its new identity with this new trend. I don't think this is something bad. It has a model of its own with all its problems. You can't criticize it because it doesn't look like a European city. This city has a structural geniousity of its own and it will use it. Istanbul will be a city where this new population will find a new language. They will transcend their local identities and create a new urban identity which is yet to be seen.[35]

As one might expect, the tensions over this new language, its redefinitions and reformulations, took their most overt political form during the municipal elections of March 1994.

The Reconquest of the City: The Municipal Elections of March 27, 1994

The municipal elections of March 1994, in which the Islamist party (Refah) won an overwhelming victory, played extensively on issues of identity and ownership. It is not possible here to provide an exhaustive analysis of these elections.[36] What follows is only a partial discussion of campaign themes pertinent to these issues and their ramifications for Beyoglu.

During the election campaign all the parties except Refah embraced the global-city project. For example, the candidate of the Social Democratic Party summarized the election in terms of a choice between a "Middle Eastern" or a "European" city. By contrast, Refah, with the slogans of "just order" and "a new world," was the only party to appeal to groups excluded from the global-city project. Refah called for the "conquest of the city the second time" by those they referred to as the "real owners of the city."

The results of the municipal elections in 1994 were a scandalous development for secularists. Refah won control of most of Turkey's big cities, including the greater municipality of Istanbul. As one sociologist put it: "Refah's most celebrated aim, 'to conquer Istanbul a second time,' has now become the nightmare of the elites."[37]

Although there was general shock in secularist circles after the elections, it was particularly difficult for secularists to believe that Refah had won in Beyoglu, Istanbul's somewhat marginal entertainment center, with its many bars, restaurants and nightclubs. The area was flooded with journalists trying to capture the "new" Beyoglu — Beyoglu "after Refah." Mainstream secular newspapers were full of articles and series with titles such as "Now What in Beyoglu?"; "Beyoglu until Refah"; and "What Next in Beyoglu?" Beyoglu was to be the most visible public space of the transformation of the city.

Of course, the victory of Refah in Beyoglu meant different things for different groups. Although it may have represented secularists' worst nightmare, a rupture with history, it had acquired several interesting meanings for Refah. Although in the Islamist narrative Beyoglu represented cosmopolitan degeneration, it also had come to symbolize many of Refah's campaign themes. And for party officials, it was a crucial opportunity to demonstrate the workings of the "Ottoman model" of government, which they defined as the coexistence of different lifestyles. This theme was highlighted in Refah's description of Beyoglu in a campaign brochure.

> Beyoglu: A Different World. Beyoglu has always been made up of different cultures, languages and faiths. Whatever their faith or language was, they all lived together in peace and harmony. Beyoglu embraced everyone.[38]

Given such a depiction, Beyoglu was the perfect place to resurrect the Ottoman model that Refah officials claimed had allowed the cosmopolitan coexistence of different groups under the tolerant Ottoman (Islamic) rule. The officials claimed that such a style of governance had enabled people from different ethnic and religious backgrounds to live in peace and harmony for centuries. Although the population of Beyoglu had now changed drastically, with most of the non-Muslim population having left, the district was still the perfect place to demonstrate this view.

Their plan for the quarter also provided Refah with a way to refute the secularists' portrayal of their party and of Islam. Refah's self-representation was that, just like Beyoglu, it could embrace everyone according to the tolerant tradition of Islam. The new mayor of Beyoglu, Nusret Bayraktar, was keen on underlining this theme. His visits to synagogues and churches in the district were highlights of the election campaign. After the elections, in an interview in the *Turkish Daily News* that stressed the theme of tolerance and pluralism, he noted, "The first thing I ordered was to clean the garbage around the Armenian church. Now how can you say that Refah is the party of Islam alone?"[39]

Contrary to many secularists' expectations, Bayraktar was also enthusiastic about preserving the architectural heritage of Beyoglu. In part, this was due to the fact that the architectural heritage of nineteenth century capitalism could be used to critique the architecture of late capitalism — most concretely represented by the city's skyscrapers. Pointing to Galata Tower, Bayraktar asked, "This tower, symbol of Beyoglu, when compared to the minarets in the old times, was qualified as bulky. Compare it now with the contemporary skyscrapers. Doesn't it look so elegant?"[40]

The post-election period manifested the different strategies by various groups to claim Beyoglu and to inscribe new meanings in the old quarter. Through analysis of these different strategies and practices, I next try to show how these discourses on the past and present identity of the city are both embedded in and played out through social practice.

CLAIMING THE QUARTER BACK

Platform for Beyoglu

It was fourteen days after the elections that the first meeting of the spontaneously formed Platform for Beyoglu took place. The room was packed with intellectuals, freelance writers, artists, political activists, architects, non-governmental organization (NGO) representatives and business owners from the district. Most of the faces were familiar from the 1980s protests and organizations against former Mayor Dalan's Tarlabaşı demolitions. Those attending were haunted by images of the new Islamist municipality, which for them meant censorship and strict control of public space.

"You know how I feel about this," one architect said. "I empathize with the Byzantines after so many years; probably this is how they felt when Fatih [Sultan Mehmet the Conqueror] conquered their city. I feel like I am under siege now."[41] His comments reflected a sentiment shared widely in the city's secularist circles. For them Istanbul was now a city "under siege."

As suggested earlier, the victory of Refah in Beyoglu meant different things for different groups. For Refah, it was a chance to revive the "Ottoman model"; for others, it marked an end to the Beyoglu they knew. But for both, Beyoglu was a key symbol. The proposed aim of Platform for Beyoglu was to discuss the implications of Refah's "victory," and consider ways of organizing against possible disruptions of their businesses — and, as they put it, their "lifestyles." One architect opened the meeting, saying:

> Beyoglu is a symbol for both sides. We have national elections in two years. Time flies and we have to come up with concrete alternatives at the micro-level. We envision this as a civil initiative which will help us to reclaim our rights in this neighborhood. What motivates us is this anxiety to preserve our lifestyles. It is resistance against any possible attack on public and cultural life. We should be organized to do this.[42]

But despite these positive words, the sense of panic and despair grew as the meeting continued. The agenda and the aim of the new organization became ever vaguer. There were different suggestions and conflicting agendas: to get back the votes — although there was no consensus as to which party to support, there being widespread disappointment with all existing left-wing parties; to define the struggle in strictly legal terms and to fight against the differential implementation of laws; and to define a mission for the group which could serve as a common denominator — one that was vaguely defined toward the end of the meeting as "to organize against the policies of Refah."

Nevizade Street

The anxiety among the secularists was well founded. One of the first acts of Beyoglu's new Mayor Bayraktar was to order the removal of restaurant tables set out on the street or on paving stones. Restaurant owners were also asked to put opaque curtains in their windows. Refah officials justified the new policies with a rhetoric of populism, arguing the street should be made more accessible to pedestrians who could not afford to eat out. One restaurant owner from the famous Nevizade Street nevertheless interpreted the event as "ideological":

> Nevizade Street is known for its restaurants which serve alcohol. On warm days the tables that we set outdoors used to be filled up by frequenters. They say they have brought this regulation because they found these tables inappropriate. But we think this is an ideological decision — they want to shut us down because we serve alcohol. Otherwise, there are many pastry shops, greengrocers and restaurants which also occupy pavement.[43]

The ban on outdoor tables provided a perfect opportunity for the Platform group to publicly organize their first demonstration. They were at Nevizade Street the next evening, organizing a sit-in, moving tables outdoors and drinking beer, *rakı*, and wine in public. They demanded their right to socialize in public space. One of the secularists' worst fears had been realized on the very first week of Refah's takeover of the municipalities. "Islamists" had begun to regulate the use of public space, pushing social life forcefully indoors.

Rain Prayer vs. The Rain Bomb

In the following days, new issues came to the fore, and new events were "culturalized" and contested, as the new municipal authorities created new sites of struggle. The issue of rain prayer was a striking example.[44]

The city of Istanbul has been plagued by chronic shortages of water, especially in the past decade, and distribution of water is among the most important duties of the municipality. During the five-year reign of the Social Democratic People's Party that preceded the March 1994 elections, the mayor of Istanbul had called for the use of rain bombs in order to fill up reservoirs close to the city.[45] By mid-April 1994, it was reported that the level of water in the reservoirs had seriously decreased. Journalists rushed to interrogate the new Refah Mayor of greater Istanbul Tayyip Erdoğan as to what he was planning to do.

When asked if he would continue to use of the rain bomb, he responded, "That [technology] was used last year and did not succeed. Today a group of our citizens are going out for a prayer for rain."[46]

"Refah Mayor Goes for Rain Prayer," was the account carried on the television news and in several newspapers the following day.[47] The mayor's critics were also quick to ridicule his refusal to use the rain bomb as proof of his "backward mentality."

The Platform group was one of those which saw the chance to organize around the issue of "rain prayer." "Claim your water," was their new campaign slogan. According to the organizer of the campaign:

> This is a great opportunity for us. What we want to do this time is to put this in nonpolitical terms. This should not look like something that we are doing against Refah — not like the Nevizade demonstration. That's why we don't have anything in our posters or pamphlets mentioning Refah or their stupid policy of rain prayer. . . . What we are doing is distributing washers for free to the inhabitants of Beyoglu, . . . It is true that there is a water shortage, but it is also true that we lose tons of water because of the dripping faucets. We'll ask people to change the washers of their faucets and save tons of water in a year. But of course the underlying message here is that we just want people to realize that there are more modern and scientific ways of dealing with the water problem, and that people should be aware of their own power over these issues rather than wait helplessly for the municipality to pray for them.[48]

Although such comments were set against the sense that a historical rupture had occurred, life went on in Beyoglu, evading such a sense and simultaneously providing a critique of it. According to Mehmet, an ex-left-wing militant who ran a bar in the district, "There is no such thing as 'post-Refah' here. That's the problem of the outsiders. Life just goes on here."[49]

For Mehmet, it was politics as usual, and he preferred to talk about the change through a rhetoric of corruption. In particular, he was furious about the activities of the Platform group. "I don't get this. It is again 'saving' Beyoglu, this time from Refah," he said.

> But these people who try to save Beyoglu do not live here and they don't know what it means to live here. That's why they see this as rupture. Let me tell you the difference between Refah and these intellectuals. You know that there are many alcoholics in this area. Intellectuals come here and tell them how bad it is to drink alcohol. Refah activists come here and they give these people money to get alcohol. That's the difference. All of these people steal, they are corrupt, but I guess Refah is just more clever, they are more practical about their politics.[50]

Orhan joined the conversation:

> Forget about all these left-wing, right-wing politics. It is true that people would be killing each other here for political reasons, as everywhere else, before the coup [referring to the political violence of the 1970s which ended with the military takeover of 1980]. But ideology does not exist

here. Whoever gives them money, these people will be on their side. Nowadays, it is the check-and-bond mafia.[51]

Orhan had grown up in Dolapdere, an area in the backstreets of Beyoglu. In the late 1970s, his father had been a left-wing militant, and had gone to prison during the military coup of 1980; in the late 1980s, his father had become an Islamist, joined a *tariqat* (religious order, banned by the state), and had been imprisoned for that, too. Orhan had started having problems with his father when his father started practicing religion.

> My father used to paint before he got into religion.... There has always been a competition between us. I taught him how to play the guitar, he taught me how to paint. Then I started drawing cartoons. He started tearing them, and one day he broke my guitar. That day I fled home. I also didn't like the way he treated my mother. My sisters accepted that.... I taught my mother how to read and write, secretly from my father. . . . This is a bizarre place. Discordance is our lifestyle. I remember when I was a kid, I found a wallet on the street. I was beaten up by my mother, but then we survived on that money for a month. On the other hand, there is also a strange moral code here, a concept of family, family values, which I could never relate to. Beyoglu cannot exist without Dolapdere. Most of the people you see there work in Beyoglu. They are the prostitutes, they are the drug dealers, they are the musicians in the most run-down nightclubs.... It is funny to hear people talking about different ideologies. There are no ideologies here.[52]

A Mosque in Taksim Square

Two months after the local elections, I joined a group of Refah supporters at one of the meeting halls owned by the municipality to hear greater Istanbul's new Mayor Tayyip Erdoğan deliver the opening speech of the week-long celebrations marking the 541st year of conquest of Istanbul by the Ottomans. Erdoğan's speech focused on the alternative reading of Istanbul's legacy and identity.

> Istanbul is a meeting place between East and West. Many different groups lived here in peace and harmony for centuries. History should be an inspiration for new models; new models of coexistence; possibility of living together in peace. We should restore our historical memory. Nowadays, we are talking about Istanbul as "the world city." But we also see that it looks more like a big village or a provincial town with all this garbage, water shortage, and air pollution. We should be realistic and revitalize this city's long-standing mission, which is the Ottoman legacy, an Islamic legacy of pluralism. We should create universal spaces where the traditional and the modern meet. But this meeting or synthesis should take into

> account [be compatible with] our historical identity. This city derives its identity from Islam: peace and love. Now this identity is lost. We should revitalize this, first by uncovering our real historical and cultural texture. We should be thinking about the real meaning of conquest, the spiritual meaning rather than the military one. It means bringing light into the dark just as Fatih [the Conqueror] did. That's what we are planning to do.[53]

As Ashworth and Tunbridge have argued, the relationship between nationalism and historical preservation and heritage has been particularly intimate.[54] Nation-states have been forged by nationalist interpretations of the past, and what is considered to be "heritage" is largely determined by these interpretations. The Islamist movement, challenging the secular nationalists' interpretation of history and their version of Istanbul's legacy, established an alternative version of national heritage apart from, and in opposition to, the official one. It was therefore not surprising that one of the fiercest political battles after the elections took place over the meaning and significance of this heritage, as provoked by the Islamist Party's proposal to build a mosque and Islamic cultural center in Taksim Square in Beyoglu. The choice of the square and the specific location of the mosque were not coincidental: the site was across from the Greek Orthodox Church, and it would also be competing with the Atatürk Cultural Center, a powerful marker of the Turkish nationalist ideology.

Mayor Erdoğan once again emphasized his party's alternative reading of Istanbul's legacy and identity in defense of the mosque project. He pointed out that the idea of the mosque was to highlight the Islamic identity of Istanbul:

> Taksim is a crucial region for tourism in Istanbul. Someone who visits this area will have a sense that he is in an Islamic city. When we gradually bring out the historical and cultural texture of our city, tourists who visit Istanbul will understand that they are in a city populated by Muslims.[55]

The messages around this mosque were complicated and embodied different possibilities. For Refah Party officials, the mosque represented the real mission and legacy of Istanbul. They hoped it could both mobilize a local constituency and appeal to a global audience. In secularist circles the plan was seen as yet another blow to the secular ideals of the Republic. According to the head of MMOB, it was "an identity battle which would be played out in Taksim."[56] Feelings were strong against the project. "The day they build a mosque on that square," said one young businessman, "I am leaving this country."[57] For the financial sector and the government, who were interested in resurrecting and marketing only a sanitized version of the Ottoman past, building a mosque in Taksim would jeopardize Turkey's chances of membership in the European Union. In addition, those involved in the tourist trade were not convinced of its appeal. According to one travel agent:

> Tourists who want to see the Islamic heritage go to the historical peninsula anyway. The real and monumental Islamic architecture is there. The mayor is just trying to please his constituency, and he is using tourism as an excuse for that. Moreover, who says that tourists come here to see the mosques. They visit Istanbul to see the Byzantine heritage or Christian heritage. When they come to Beyoglu, they visit some churches and Galata Tower, that's it.[58]

This controversy took place at the same time that Refah officials were declaring their intent to demolish the Byzantine city walls surrounding the historical peninsula. The motivation for such a demolition was put rather bluntly by one official: "We don't want a Byzantine Istanbul."[59] The proposal led to a series of debates both within and without Islamist circles. Recently, historic preservation has been framed as a sign of tolerance, where everyone is expected to "join the global drive to preserve the great architecture of all periods and civilizations."[60] But many more issues were evident in Istanbul at the time. Whose heritage was going to be preserved and for whom? If Refah insisted on demolishing the walls, how would it sustain its image of multiculturalism and tolerance (at least that image promoted by its moderate factions)? And how would demolishing part of the Byzantine heritage effect tourism in a city that catered to Western tourists, most of whom visited it for its Christian heritage?

The proposal for such a demolition was made even more problematic by its timing. Turkey was then applying for Customs Union with Europe, and the central government was trying to promote a Western-looking, secular image. Prime Minister Çiller flew to Istanbul the following week, where she held a press conference in front of the city walls to advocate the historical value of such heritage. As a result of such pressure, the proposal was eventually withdrawn.

Galata Festival

One of the most outspoken activists for Beyoglu was Metin. As one of the leaders of what he called the "urban renaissance movement" in Galata, he had been organizing the Galata Festivals for more than a year. Sitting in his restored office/house in Galata (a building originally built for a Maltese merchant in the nineteenth century and then used as a British police station during the World War I occupation of Istanbul), he explained his view of what the district.

> Although I am an architect myself, I am fed up with the architects. They live in such an abstract world. For every line you draw, there is a philosophy behind it. The thing you draw on paper is 10 percent of the whole thing. The real issue is the people; how will you implement those plans? . . . Actually, forget about the plans — who cares about them anyway? It is the real estate developers who have the power these days.

Pouring a glass of wine, which he had gotten from the Catholic Church across the street, he continued:

> Pera was the vineyards in the early periods of Galata settlement. Pera vineyards. In 1838 Galata becomes the center of foreign capital which flows to the empire as the result of Ottoman-British Commercial Treaty. Pera is the place where you see the first apartment buildings. They were built in the nineteenth century by the architects who were familiar with the European lifestyle. When the Ottoman, Middle Eastern, and Asian lifestyles were added to this, we see an emergence of a cultural synthesis here. Beyoglu, especially Tarlabaşı, has been the open-air museum of this hybrid architecture — well, until the 1980s demolitions, of course. There is a continuity in all of these things. We all agree that we have to do something about Beyoglu, make it more livable, and preserve these historical buildings. But the problem is to make the inhabitants of this area conscious of the value of these buildings. . . . That's what we are planning to do through these Galata festivals. This is part of the urban renaissance movement here, We are the neo-Levantines.[61]

Later, on May 21, 1994, at the first Galata Festival after the Refah Party victory in the municipal elections, tensions were high. Metin had been keen on inviting Mayor Bayraktar to deliver the opening speech, as had been done with previous mayors. This had provoked a split with the Platform group, one of the sponsors of the festival, because it did not want to associate with the new mayor. But Metin did not give in to their demands, and he invited Bayraktar. The Platform group thereupon protested the festival, refusing to attend, and asked Metin not to associate with it any longer.

When I mentioned the festival to her, the owner of one of the pastry shops in the area started complaining about the Chamber of Architects (MMOB). "If you ask me," she said, "these buildings are just dirt. MMOB insists that they should be preserved. Preserving this dirt, I don't get it. I don't see what is historical and what is valuable about them."[62]

The owner of the coffeehouse near the Galata Tower had similar complaints. "Yes, a lot of students come here, they tell us that these are historical buildings, but what is the use? That's what I want to know." His coffee shop was crowded with students of architecture who were in the neighborhood to document and measure the buildings. He told me that this was routine: students would come, take pictures of the buildings, draw, and chat with him about the buildings' "historical value." Although he appreciated the students' efforts, he had a more utilitarian approach. "Tourists come here, but they just go up the tower, that's it. There is no place where they can spend money. These buildings have no use for us."[63]

Ali, who had been working as a real estate agent in Galata, and was also the local administrator of the two districts of Galata, explained things this way:

> My father arrived here in the mid-1950s. He started to work as an apprentice to a Greek plumber. Then he bought this shop. Later on, his relatives and friends started moving here. There were only two Turkish families then — the rest were Jewish. Later keepers here were from Erzincan, and they started buying the houses from the Jews who were leaving. The tendency has always been to vote for the center-right parties here, but more and more it is Refah now. SHP [the Social Democratic Party] did not do anything here. They would not even meet with us. These new ones claim that they will have regular meetings and so on. I really like Metin, he has good intentions, but I don't think festivals like this will work here. This place is not really residential anymore — it is mostly workshops.
>
> But there is something interesting happening here for the last six years or so — there are some journalists, architects, painters, artists moving here. I even have a client who moved to the U.S. couple of years ago. Her mother, who still lives in Istanbul, came here the other day to buy an apartment for her daughter. Well, apparently they realized that you cannot "save" Beyoglu from afar — you have to live here. First it starts with your own apartment, then the building, then the neighborhood. You can't do it from afar.[64]

CONCLUDING REMARKS

The different positions, alliances and strategies that I have here described bring into focus the complex meanings that heritage and historical preservation/revitalization acquire. The struggle over Beyoglu brings up the issues of "who we are/who we were." What it means to "Turkish," "European," "modern" — what becomes "local" and "global" — are all negotiated and contested around this built environment. These struggles are not only shaped by the (re)imaginings of the past, but also by the forces of global politics and local identities in the present. Simultaneous (re)workings of local and global converge around the discursive and material practices that concern these buildings. Sometimes these buildings are "Turkish," sometimes "European," and sometimes they are "Levantine"; sometimes they are "our own," and sometimes they are "other."

As Stuart Hall has suggested, although the "cultural identity" of a city has become essential material and symbolic capital in the new race between cities, these identities are "far from being eternally fixed in some essentialized past, they are subject to the continual play of history, culture and power."[65]

Cultural heritage, preservation and conservation become contested domains through which the past, present and future are (re)worked and (re)formulated. Globalization is inscribed within particular localities and is reworked within particular social, cultural and historical contexts. Within this framework, heritage and the politics of the past take on a very different meaning, and which past to preserve and market, and whom to market it to, become political questions. Cities, as physical embodiments of histories, become crucial sites where different claims to the past are formulated and contested. This formulation provides opportunities to explore the reworkings of modernity and globalization through the politics of heritage in urban contexts.[66]

NOTES

This chapter has been presented in various forms at seminars at the Department of Anthropology at the University of California at Berkeley and at the Middle East History and Theory Conference at the University of Chicago. I am grateful to the participants of these seminars and the conference for their comments on earlier versions.

1. I use the names Pera and Beyoglu interchangeably to refer to the same area. Although these terms connote different meanings and histories, space limitations preclude any discussion here of the relevant politics of naming and boundaries.
2. AESOP tour pamphlet.
3. As recorded by the author in an interview, August 25, 1994.
4. S. Errington, "Fragile Traditions and Contested Meaning," *Public Culture* 1, 1989, pp.49-59; R. Handler, "On Having Culture: Nationalism and the Preservation of Quebec's *Patrimoine*," in G.W. Stocking (ed.), *Objects and Others: Essays on Museums and Material Culture*, Madison, Wisconsin, University of Wisconsin Press, 1985; M. Herzfeld, *Ours Once More: Folklore, Ideology, and the Making of Modern Greece*, Austin, University of Texas Press, 1982; *A Place in History: Social and Monumental Time in a Cretan Town*, Princeton, Princeton University Press, 1991; and R. Johnston et al. (eds.), *Making Histories: Studies in History-Writing and Politics*, London, Hutchinson, 1982.
5. F. Elliot, *Diary of an Idle Woman in Constantinople*, Leipzig, Collection of British Authors, Tauchnitz Edition, 1893, p.12.
6. S.T. Rosenthal, *The Politics of Dependency: Urban Reform in Istanbul*, Westport, CT, Greenwood Press, 1980.
7. This is a term used by the Ottomans to refer to Westerners.
8. Z. Çelik, *The Remaking of Istanbul: Portrait of an Ottoman City in the Nineteenth Century*, Seattle and London, University of Washington Press, 1986; and Rosenthal, *Politics of Dependency.*
9. O. Arkan, *Beyoglu: Kısa Geçmişi, Argosu* (*Beyoğlu: A Brief History and Its Slang*), İstanbul, İletişim Yayınları, 1993.
10. P. Mansel, "Plans to Make Istanbul a Center of Culture," *Financial Times*, supplement on "Istanbul and the Olympics," June 24, 1993, p.3.
11. K. Robins and A. Aksoy, "Istanbul Rising: Returning the Repressed to Urban Culture," *European Urban and Regional Studies* 2, vol.3, 1995, p.6.
12. D. Harvey, *The Condition of Postmodernity: An Enquiry into the Origins of Culture Change*, Cambridge, Basil Blackwell, 1989, p.271.
13. Ç. Keyder and A. Öncü, *Istanbul and the Concept of World Cities*, Istanbul, Friedrich Ebert Foundation, 1993, p.29.
14. A. Üçok, "Hepsini Yıkacağız," interview with B. Dalan in *Şehir*, March 1, 1987, pp.78-79.
15. As recorded by the author in an interview, 1987.

16. Ibid. The reference to September 6th and 7th refer to events that took place in the late 1950s, when the shops and houses of mostly Greek minorities were looted and burned in response to the news that Ataturk's house in Thessaloniki had been bombed. The Greek minority was forced to leave the country. Although these events are rather recent, they are in no way part of official history, and they are hardly part of popular historical narration. Many people become very irritated when asked about these events.

17. As recorded by the author in an interview, 1987.

18. Ibid.

19. Arkan, *Beyoglu: Kısa Geçmişi, Argosu*, p.27.

20. Keyder and Öncü, "Istanbul and the Concept of World Cities."

21. Name of the harbor region of the area.

22. A more common term used for the description of this change in the urban scene is *arabesk*. For a discussion of *arabesk* culture in Turkey in the late 1980s, see M. Stokes, *The Arabesk Debate: Music and Musicians in Modern Turkey*, Oxford, Clarendon Press, 1992.

23. As recorded by the author in an interview, February 1994. One of the characteristic foods of southern and southeastern Turkey and of Arab countries, *lahmacun*, became popular in cities like Istanbul with the 1950s rural migration to the cities. It is often mentioned by members of the middle and upper classes as a way of expressing their resentment of migrants.

24. The name that was used mainly in the nineteenth century to refer to the main street in the area. The name was changed to İstiklâl Street (Street of Independence) with the establishment of the Turkish Republic.

25. As recorded by the author in an interview, 1988.

26. Ibid.

27. As recorded by the author in an interview, 1987.

28. Ç. Gülersoy, "Doğu'ya Giden Teknede Batı'ya Koşuyoruz," interview in *Şehir*, July 17, 1987, p.46.

29. As recorded by the author in an interview, February 1994.

30. Interview with the author, March 1994.

31. E. Ihsanoğlu, *İstanbul: Geçmişe Bir Bakış* (*İstanbul: A Glimpse into the Past*), İstanbul, Research Center for Islamic History, Art and Culture, 1992, pp.6-7.

32. D. Dursun, "Which İstanbul?" *Nehir*, March 1995, p.15.

33. F. Müftüoğlu, "İstanbul'un Perişan Hali ve Islah Çalısmaları," *Nehir*, March 1995, p.9.

34. T. Bora, "Fatih'in İstanbul'u: İslâm Şehri ile Dünya Şehri Arasında Islâmcıların İstanbul Rüyası," *Birikim* 76, August 1995, pp.44-53.

35. TV Interview, March 1994.

36. For a more comprehensive analysis of the election results, see R. Çakır, *Ne Şeriat Ne Demokrasi*, İstanbul, Metiş Yayınları, 1994; and E. Kalaycıoğlu, "Elections and Party Preferences in Turkey: Changes and Continuities in the 1990s," *Comparative Political Studies* 4, 1994, pp.35-48.

37. T. Bora, "İstanbul'un Seçimi," *Birikim* 56, 1994, p.4.

38. Refah campaign brochure.

39. N. Bayraktar, "Our Success Will Be Seen," interview in *Turkish Daily News*, April 13, 1994, p.B1.

40. TV Interview, March 1994.

41. As recorded by the author in an interview, April 11, 1994.

42. Ibid.

43. As recorded by the author in an interview, April 1994.

44. Many of these issues were publicly debated and had a deep imprint on the popular memory of inhabitants of Istanbul. That is why they are referred to as "the issue of" such and such.

45. Rain bombs have been used for cloud seeding when there has not been enough rain in the city.

46. As recorded by the author.

47. Ibid.

48. As recorded by the author in an interview, May 1994.

49. Ibid.

50. Ibid.

51. Ibid.

52. Ibid.

53. June 1994 opening speech.

54. G.J. Ashworth and J.E. Tunbridge, *The Tourist-Historic City*, London and New York, Belhaven Press, 1990.

55. T. Erdogan, interview in *Yeni Zemin*, March 30, 1994.

56. As recorded by the author in an interview, July 1994.

57. Ibid.

58. Ibid.

59. TV interview, July 1994.

60. Herzfeld, *A Place in History*, p.67.

61. As recorded by the author in an interview, February 1994.

62. As recorded by the author in an interview, May 1994.

63. Ibid.

64. Ibid.

65. S. Hall, "Cultural Identity and Cinematic Representation," *Framework* 36, 1989, p.70.

66. A. Pred and M.J. Watts, *Reworking Modernity: Capitalisms and Symbolic Discontent*, New Brunswick, NJ, Rutgers University Press, 1992.

7

Learning from Chinatown: The Search for a Modern Chinese Architectural Identity, 1911–1998

Anne-Marie Broudehoux

Earlier this century American city planner Edmund Bacon described Beijing as "possibly the greatest single work of man on the face of the earth," and architectural historian Steen Eiler Ramussen praised the city as "one of the wonders of the World, . . . a unique monument, the culmination of a great civilization."[1] Today many local and foreign observers fear that this vision of old Beijing has all but vanished, that Beijing is no longer "Beijing" — or worse, that it has been turned into a Chinatown.[2]

Since the mid-1980s, the city of Beijing has been the object of important restructuring through redevelopment to accommodate modernization and attract foreign investment and visitors. Modern infrastructure and high-rise icons of progress have replaced entire sections of the old city. One of the most striking features of recent transformations has been the apparent re-sinicization of Beijing's image, especially with the development of a new architectural form which makes profuse use of features borrowed from China's architectural traditions. Most new buildings recently erected in the capital integrate Postmodern architecture and diverse regional and historical decorative elements, of which the most common is the pastiche glazed-tile roof. Such developments have provoked both praise and criticism in China and abroad, and have revived a century-old debate about the definition of an architectural expression of modern Chinese identity which would allow for the modernization of the age-old capital city without altering its uniquely Chinese character.

The fear of losing the "essence" of old Beijing — often perceived as the physical embodiment of Chinese civilization — is nothing new. Throughout this century, foreign and local commentators alike have perceived modernization as a threat, not only menacing the identity of this ancient city, but also the whole of Chinese culture as well. In the late 1920s, the British and American authors of *In Search of Old Peking* deplored early modernization efforts in the Chinese capital as a "serious ill-omen for the artistic and cultural future of the country as a whole."[3] Such fears were often based on a perception of Beijing — and of Chinese identity by extension — as unchanging, eternally frozen in the past and isolated from foreign influences.

In reality, throughout Chinese history, foreign cultures have infiltrated the country's borders and enriched its cultural traditions. In 1919, Juliet Bredon rightly wrote, "The history of Peking is the history of China in miniature. The town, like the country, has shown the same power of taking fresh masters and absorbing them."[4] The development of modern architecture in China is no exception, and this chapter demonstrates that the century-long search for an architectural expression of modern Chinese identity has been the product of a convergence of agendas from actors both inside and outside Chinese society. Thus, to understand the significance of recent architectural developments and image-making efforts in the Chinese capital, it is necessary to review the historical development of modern architecture in China.

GLOBALIZATION AND THE CULTURAL HYBRIDIZATION OF IDENTITIES

The late twentieth century has been marked by important worldwide economic, institutional and cultural transformations. The unprecedented scale and rate at which such transformations have taken place have been unsettling for people around the globe. Increased exposure to alternative ideas and modes of living — through the media, the spread of consumer culture, or through direct contact with "Others" at home or abroad — has led to increased questioning about the nature of local distinctiveness, and has triggered a reevaluation of local values. Globalization has thus been accompanied by the exacerbation of societal and ethnic self-consciousness and the resurgence of expressions of collective identity.

At the heart of current discussions of global transformations, and closely tied to issues of identity, has been a growing fear of cultural homogenization. Globalization — often equated with Americanization — is now increasingly blamed for erasing local particularisms. Yet such allegations rest on a misunderstanding of the nature of identity — which may be mistakenly presented as stable, permanent and coherent. By failing to recognize the complex, multiple, and ever-changing nature of identity, proponents of "homogenization theory" deny local agents a role in transforming or appropriating extralocal influences to serve their own interests.[5]

Hybridization, a nineteenth-century concept which has been recently reactivated as a key issue for cultural debate, seems more appropriate than homogenization in describing worldwide cultural change. Although the term has been narrowly defined as the organic grafting of diversity into singularity, hybridization is better understood as the fusion of different elements, resulting in the creation of new entities. Cultural globalization should thus be envisioned much more as the increased interfusion of diverse particularities than as the diffusion of a single form.[6]

According to Robert Young, the concept of hybridity also suggests the impossibility of essentialist notions of cultural identity. Arguments about homogenization fail to consider that all cultural practices have historically been constructed by the incorporation of extralocal influences, and have in turn affected practices beyond their own boundaries. As elements from various cultural groups are brought into new societies, they rapidly become indigenized in one way or the other. Today, new, transitional or hybrid identities are emerging where civilizations intersect, and concepts of homogenous national cultures or organic ethnic communities are in a profound process of redefinition. Although globalization may actually bring greater diversity and trigger renewed interest in local distinctiveness, it may also lead to a dangerous retreat into the security of nationalism and fundamentalism.[7]

To grasp how these new identities are being constituted at the dawn of the present century, it is necessary to examine the origins and development of such hybridization processes and to understand the way new cultural meaning is constructed and transformed over time. This chapter thus examines the genesis, development and evolution of modern Chinese architecture, especially the search for an architectural expression of Chinese collective identity, from its origins as an Orientalist import to its appropriation as a national symbol.

The nineteenth century, with the development of new transportation, communication and production technologies that facilitated the spread of imperialist capitalism, was one of the most intense periods of cultural mixing in world history. In China it marked the beginning of a series of transformations which forever changed the face of the Middle Kingdom. Ever since China was forced to open its doors to Western powers with the Treaty of Nanjing following the First Opium War in 1842, the Chinese have faced the dilemma of choosing between tradition and modernity — often perceived as a choice between Eastern and Western influences. From the late nineteenth century Confucian reforms under the slogan "Chinese learning for Essence, Western learning for Use," to the more recent search for a sinicization of Marxism and the definition of "Socialism with Chinese characteristics," the question of the place of Chinese traditional culture within modern China has been central to the definition of Chinese collective identity. Such struggle to redefine the image of the Chinese nation has been reflected in the search for an appropriate built form which can represent China to the world — a search which was initiated at the beginning of this century, and which is still going on. Over the years this search has often taken the form of an ideological battlefield, as architecture has become a tool to serve the interests of those parties seeking economic power, political control, or doctrinal rectitude.

THE GENESIS OF MODERN CHINESE ARCHITECTURE: REINVENTING TRADITION

It was only in the late 1920s that the first Chinese trained abroad as architects returned to China and established the first schools of architecture. Modern building construction in China's treaty ports in the late nineteenth and early twentieth centuries was thus monopolized by foreign architects — first British, and later Americans and continental Europeans. It was these outsiders who introduced new building types to China, such as offices, hotels, clubs, residences and municipal infrastructure, and who reshaped the land in the concessions governed by foreign powers to match "modern" needs and ideals.[8]

Early foreign construction in China displayed many eclectic European styles, but the most popular was a brand of Classical revivalism which was christened by Western expatriates in China the Compra-doric style.[9] Occasionally, late nineteenth century taste for the picturesque also led to the use of indigenous ornamental elements such as flaring eaves, bracketing, and gilded signboards — particularly on storefronts. But the first explicit attempts at creating a modern architectural style with Chinese characteristics were made by foreign architects, mainly Americans, commissioned by missionary organizations. Missionaries sought to give a local color to their colleges and hospitals by introducing Chinese details to buildings whose basic layout and structure embodied the most modern innovations. The new style, blending Western technique and Chinese ornament, came to be known as the "Chinese Renaissance" or the "Chinese style."

One of the first advocates of the Chinese style was Harry Hussey, a Canadian architect who practiced periodically in China after 1911. Since early childhood Hussey had been attracted by the mysteries of the Orient. One of his most famous designs was for the campus of the Peking Union Medical College, which was commissioned by the Rockefeller Foundation in 1916. Hussey recommended that the Chinese style of architecture be used for the college out of concern for the preservation of the old city's character and to counter the proliferation of Compra-doric style buildings in Peking **(Figure 7.1)**.[10] Such a stylistic choice (for an institution which, in the words of John D. Rockefeller Sr., was to exert "an ever-widening influence for the promotion of the physical, mental and spiritual well being of the Chinese nation"[11]) was also justified as a way to make the college "something not foreign to China's best ideals and aspirations, but an organism which will become part of a developing Chinese civilization."[12]

Such an early venture into the development of an architectural hybrid was not without controversy. The decision to use the Chinese style for the architecture of the Medical College was opposed by missionaries on the school board who deplored the loss of a good opportunity for showcasing the superiority of American architecture. The architecture of the campus was also derided by many Chinese observers for its lack of authenticity in the use of traditional elements. Indeed, despite Hussey's claims to have been directly inspired by traditional Chinese construction systems, all the glazed-tile roofs at the college were supported by reinforced concrete rather than a conventional wooden post-and-lintel structure. Although Hussey asserted that he had worked in close collaboration with architectural historian Zhu Qijian to ensure fidelity to architectural precedents, many of the college's buildings featured elements foreign to Chinese tradition.[13]

Despite Hussey's early involvement in the development of the Chinese style, it is the American architect Henry Killam Murphy who is generally taken to be its primary instigator, and the main catalyst for the fusion of traditional

Figure 7.1. The Peking Union Medical College.

Chinese architecture and modern technique. Murphy's early specialization in school and campus design in the United States was instrumental in his being chosen in 1913 to design several missionary colleges in China — where he remained active until 1935. In the 1920s, Murphy elaborated the concept of an "adaptive Chinese Renaissance," which he compared with the Greek and Roman revivals of the European Renaissance.[14]

Murphy despised previous adaptations of the Chinese style for being little more than the crowning of modern buildings (whose core structures remained essentially foreign) with Chinese roofs. He deplored the fact that most foreign architects treated Chinese architecture either as an archeological artifact or as a source of ornament to be applied to modern plans and elevations. In contrast, Murphy professed seeing the architecture of China as a living thing, and he hoped to resuscitate certain forms of traditional architecture by applying them to contemporary problems.[15]

Despite such high aims, Murphy's method retained much of the facadist approach he had been using in his North American campus design, where efforts were first concentrated on the plan, with elevations later executed in a variety of styles. Although he emphasized the importance of giving the lower portion of a building a more "Chinese" look — with the help of latticework and the "simulation" of lacquered Chinese columns — Murphy generally limited the display of Chinese features to the exterior of his buildings. The architectural accuracy of his buildings was also questionable: studies of local and regional architectural traditions were rare, and so were consultations with local master builders. Moreover, in the early years of his work in China, plans were generally drawn abroad by people who had never set foot in Asia, using published photographs of the Forbidden City and measured drawings of ancient imperial architecture as a source of inspiration.[16]

Despite having been heralded as the initiators of the Chinese Renaissance movement, people like Murphy and Hussey did not operate in a vacuum. Many factors influenced the genesis of this new architectural form in early twentieth century China, factors that had less to do with the personal agendas of a handful of foreign architects, and much more to do with a convergence of circumstances specific to the historical moment. Local conditions, including the rise of Chinese nationalism as well as the missionary nature of the institutions which both commissioned and were to be housed in the projects, played a part in stylistic decisions. External influences, such as global ideological currents, international architectural trends, and general attitudes toward "exotic" cultures, also acted on the development of this new architecture, and the Western imperialist ideology of the time, including its construction of Orientality, were part of the specific historical context within which early modern architectural developments in China took place.

THE ORIENTALIZATION OF CHINESE ARCHITECTURE: THE MISSIONARY YEARS

One factor that contributed to the construction of a global image of China was the series of "world's exhibitions" which proliferated in Western nations at the turn of the century. World's fairs are wellknown sites for the construction and diffusion of stereotypes about faraway peoples.[17] European and North American world's fairs of the late nineteenth and early twentieth centuries were often marked by the racist tendencies of Social Darwinism, where progress was considered to be a phenomenon occurring only in a few nations of European origin. Aside from displays of technological innovation, the fairs often contained ethnic exhibits which were not run by the countries concerned, and in which non-Europeans were portrayed as exotic, primitive people with little hope of improvement.[18]

At the world's fairs, the Chinese were generally portrayed as an unprogressive people, deprived of imagination and incapable of novelty, but whose peculiar ways incited both fascination and curiosity. For example, at Philadelphia's Centennial Exposition of 1876, the local *Public Record* offered this portrait of the exotic Chinese: "Cut off from the rest of the world by the great wall, and isolated behind her old feeling of distrust and apathy towards the peoples of Europe, the old empire of China has received but little benefit from western civilization and advancement, and as a consequence, makes a display at the exposition which is curious, rich, and rare."[19] Chinese people were seen as cunning and uncivilized, frozen in an ancient past. Their unyielding character was at times used to legitimize Western imperialism in China, and their backward ways made it Europeans' duty to go and enlighten them.[20]

Such attitudes also permeated the work of foreign architects practicing in China at the turn of the century. Despite his genuine admiration for classical Chinese architecture, and his real desire to help China modernize, Murphy failed to see the contradictions inherent in his personal *mission civilisatrice*: "showing the Chinese the possibilities of their own architecture."[21] Murphy deeply believed that China needed the helping and knowledgeable hand of the West to find its way to enlightenment. Thus, in 1928 he wrote:

> It has long been an article of my architectural faith that in every country having its own distinctive architectural development, an architect — whether himself of that country or from outside — should, in building, preserve that style. The fact that the architecture of China is so very distinctive of the country makes it all the more important that this principle be observed.[22]

Murphy's first commission in China was for Yale University's new campus in Changsha, in the southern province of Hunan, where the American institution had established an educational center in 1903 at the invitation of a group of Christian missions[23] **(Figure 7.2)**. For the construction of this new campus, the Yale-in-China Committee had required that "every possible step [should] be taken to conserve the architectural heritage of old China, blending it with the structural requirements that the modern world makes today."[24] It was believed that this architectural style would have educational value, in "showing the Chinese the possibilities of preserving their architectural heritage."

Murphy's interpretation and use of Chinese tradition were clearly defined in Western terms. His concept of authenticity was highly flexible and could be adapted to meet circumstantial requirements, suit his design ideas, and fit Occidental taste. For example, in the design of the Yale-in-China campus, Murphy chose to use the Ming Dynasty architecture of the Forbidden City — located more than 1,300 km. north of Changsha, in Beijing — as a source of inspiration. (In general, the simpler,

Figure 7.2. Campus elevation of the library, Yale-in-China, Changsha.

more restrained lines of northern Chinese architecture were more appealing to Western sensibility than the local southern architecture, with its extravagant ornamentation and widely flaring eaves.) Traditional architecture was also adapted to fit conventional institutional plans, and was sometimes transformed by adding extra stories and dormer windows to gain extra roof space. Such unfettered recasting of historical precedents raises questions about the sincerity of Murphy's deference for Chinese tradition, making it appear little more than skin deep. In fact, his designs maintained only an accessory role for Chinese architecture — using it merely as a source of decorative elements to create "picturesque effects." Meanwhile, he showed more interest in simulating the overall "Chinese" character than in retaining the structural logic and integrity of the local building tradition.[25]

Murphy's architecture was also shaped by his own romantic ideas about Chineseness. Murphy believed in the existence of an "authentic Chinese culture," which he found in the picturesque canals, the little arched bridges, the wall and moat, the narrow streets, and the fast-moving sedan chairs he encountered during a visit to Suzhou.[26] His image of the "real" China was thus closer to that of the imperial past than the Republican present. His concern for keeping intact what he saw as the essence of Chinese identity may thus have been driven by a fear of losing this sanctuary of "Otherness," essential to the constitution of Western identity, and thus to his own self-definition.[27]

Missionary commissioners held a similarly essentialist conceptualization of Chinese culture. This is particularly visible in their calling for the Chinese style to be used only for specific buildings on their campuses. At Yale-in-China, buildings whose function was of a scientific nature or that were to house a predominantly American population — such as the hospital and the professors' residences — followed a distinctively American Colonial style. On the other hand, the library, chapel, school buildings and dormitories, where students would spend the greatest part of their time, were built in the Chinese style. Similarly, at Tsinghua University in Beijing, also designed by Murphy, the administration considered that "buildings which were to house the latest Western ideas should show the latest ideas of that civilization in terms of architecture."[28] This suggests a certain conception of the West as the realm of progress and science, and of the Orient as beholden to culture, spirituality and tradition.

Also implicit in the decision to use the Chinese style for the architecture of the Yale-in-China campus in Changsha was Yale's desire to downplay its American roots and make its missionary presence less visible.[29] This was especially important in light of the anti-foreign riots which had erupted in Changsha in 1911, three years before the construction of the new campus, and which had been anti-missionary in nature. The use of indigenous architecture was therefore seen as a way to facilitate the infiltration of Christian values under a familiar guise, and to make the foreign presence less obtrusive in the face of the threat of social unrest.

Missionary commissioners in China were not the only ones to use architecture to attain "psychic pacification" at the turn of the century.[30] Driven by similar motives, many colonial architects in the Orient also attempted to develop a modern architectural style that incorporated local traditions. Luytens' use of the Indo-Saracenic style in New Delhi's government buildings in 1911, Lyautey and Prost's neo-Moroccan designs in Rabat and Casablanca between 1910 and 1920, and Hermann Jansen's 1923 designs for Ankara, which incorporated neo-Ottoman features, all blended foreign and local features in an attempt to pacify the natives and stabilize colonial domination.[31]

Hussey and Murphy's ventures into the Chinese style were shaped by this general architectural climate which promoted exotic developments. During the first years of this century, North American architects also remained more conservative than their European counterparts, and were slow to embrace the Modern Movement then burgeoning in Europe. In particular, North Americans retained a nineteenth century taste for the exotic and the picturesque, influenced by late Victorian sensibilities and fueled by the Oriental architectural fantasies often found at the world exhibitions. American architects had a further propensity for so-called period styles — where "style" was understood to be a visual language of surface decorative details that could be applied independent of structural or organizational components. One final influence was the Arts and Crafts movement, a product of late Victorian Britain, which flourished in North America at the turn of the century, and which promoted the idea that every country should have an architecture that reflected its own particular history.[32] This movement may have contributed to the mind-set that drove North American architects like Hussey and Murphy to develop the Chinese style of architecture in China.

It is worthwhile at this point to note that another outcome of this North American architectural and ideological climate was the hybrid architecture developed to house Chinese immigrants in so-called Chinatowns of the same period. As discussed by Christopher Yip, early Chinese migrants to the United States did not initially import any building types from their native places.[33] The first Chinese Americans lived in existing tenement houses, which were undifferentiated from other local buildings except for occasional signs in Chinese ideograms or decorative lanterns. It was only after the complete destruction of San Francisco's Chinatown by fire in 1906 that this settlement, the first and largest in North America, was rebuilt with an explicit intent on the part of municipal authorities, landowners and Chinese merchants to exploit the exotic qualities of Chinese architectural details and ornaments. To capitalize on what had recently become the city's principal tourist attraction, city officials talked of building "an oriental city . . . with the features outwardly characteristic of a Chinese city, with its pagodas, its temples and its lantern porticoes."[34] The San Francisco Real Estate Board also suggested that all property owners in Chinatown build new structures with an "oriental and artistic" appearance.[35]

"Chinatown," as it was now known, was therefore designed by American architects who had been hired by local landlords and Chinese merchants. For inspiration these designers turned to published images of monumental Chinese architecture, especially of the Forbidden City in Beijing — which, ironically, had little to do with the vernacular forms of rural southern China, where most Chinese in North America had originated. In general these designers' attempts to give buildings an

Oriental flavor were confined to facade treatment, roof overhangs, and the occasional placement of pagoda-like towers at the corners, whereas the structures themselves corresponded to conventional local building types.[36] This stereotyped image was soon adopted and reproduced by Chinese merchants and Chinese benevolent associations to gain visibility in their communities and to attract tourism. Over time this image spread to other Chinatowns and became consecrated as the physical manifestation of the presence of ethnic Chinese communities around the world.

The obvious similarities between Chinese Renaissance architecture and Chinatown architecture are far from fortuitous. These two architectural expressions were shaped by similar ideologies, based on the idea of "Chinese" as "Other," and producing the "Oriental" in the image of the West. More interestingly, these Orientalist constructions were both subsequently recuperated by ethnic Chinese and reproduced to serve their own economic and representational purposes.

REPUBLICAN REINTERPRETATIONS: THE NATIONALIZATION OF THE CHINESE STYLE

In 1921, the first wave of Chinese students trained as architects abroad started returning to China, many from the United States where they had enjoyed scholarships from the Boxer Indemnity funds.[37] Several of these young architects, influenced by their education abroad in the Beaux Arts tradition, ignored Chinese architecture as a tradition worthy of study.[38] Others resisted the irony of reviving imperial architecture only a few years after the overthrow of the Qing Dynasty. A small — but influential — group, some of whom had worked as apprentices for foreign proponents of the Chinese style, became interested in developing a national architecture inspired by the traditions of the past. They also sought to distance their work from previous imitations of Chinese architecture by foreign architects, and they aimed instead to develop what they saw as a more authentic Chinese approach.[39]

This first generation of architects was unanimous in its criticism of the Chinese architectural Renaissance, which in its view lacked a thorough understanding of Chinese architectural systems, and merely resulted in anachronism and architectural fallacy.[40] Many were irritated at Westerners' clumsy efforts to "think" Chinese, and at the way *chinoiseries* such as the tiled roof had become eye-catching emblems of Chinese identity. Critics insisted that the heavy tiled roof was not the essence of Chinese architecture, nor was the Forbidden City the only precedent.[41] Special efforts were thus made within the architectural community to distinguish between the production of foreign and local architects. Neotraditional buildings designed by foreigners were labeled "Chinese style"; similar work by Chinese professionals was identified as "Indigenous style."[42] Such designations were very important for Chinese architects, as each category carried different connotations in terms of authenticity. In reality, both styles remained very similar in their basic characteristics, and both approaches rarely ventured beyond stylistic and formal explorations. Few buildings attempted to develop a more spatial understanding of Chinese architecture.

In 1919, William H. Chaund, a virtually unknown young man from Canton who had studied architecture in Chicago from 1913 to 1917, published a passionate manifesto on the problem of developing a modern architectural style fit for China. This style, he claimed, should not slavishly imitate Western achievements, but would rather "express the . . . tradition, national spirit and dominating ideal of the period."[43] Chaund thus made one of the first calls for the development of an architecture that he called "national in character and joyously Chinese in spirit."[44]

Chaund was not alone in his conviction that China should develop its own brand of modern architecture based on local heritage. Fan Wenzhao, who had in the past collaborated with Murphy on some Chinese commissions, also came to be known as a pioneer in adapting Chinese classical architecture to modern requirements.[45] However, the event that is generally recognized as marking the origins of the Indigenous style in China was Lu Yanzhi's winning entry in the 1925 competition for the Sun Yatsen Mausoleum, built in Nanjing between 1926 and 1929 **(Figure 7.3)**. Lu had served as a draftsman for Murphy in his New York office between 1918 and 1922, where he had learned the principles of the Chinese Renaissance while working on several of the architect's most famous China projects. As one of the first national monuments designed by a Chinese architect, the Mausoleum, with its simple lines, white concrete walls, and blue tiled roofs became the symbol of national architectural production in China during the Republican era (1911–1937).[46]

Figure 7.3. Entrance to the Sun Yatsen Mausoleum, Nanjing.

The rise of Chinese nationalism in the first decades of the twentieth century provided a political and intellectual climate favorable to the search for a uniquely Chinese modern architecture. The intellectual May Fourth Movement of 1919 was itself deeply rooted in the ideal of finding a way to accept foreign technology while retaining a Chinese identity. Despite the iconoclastic anti-traditionalist tendencies of the May Fourth Movement and its embracing of foreign notions of rationality and modernity, its search for the essence of the "Chinese character" was accompanied by a renewed interest in the preservation of the Confucian past and by attempts to adapt this past to serve the future of China. The paradox of using imperial architectural traditions as symbols of Republican China was thus representative of the contradiction inherent to early Chinese nationalist movements.

Early Republican architectural developments in China were also influenced by events taking place in Japan. At the time of Murphy's first visit to Asia, discussions about the elaboration of a modern architecture with local characteristics were already common in Japan, where the architectural profession was well developed. Japanese architects were struggling to find new ways to maintain independence in the face of Western imperialism, and to embrace modernity without being dominated by Western ideas. Most Japanese were vigorously opposed to the use of a so-called "modified Japanese style" by foreign architects who could not grasp the subtleties of Japanese architecture.[47] Architectural developments in Japan, where many Chinese architects were trained in the 1920s, thus strongly influenced the progression of architecture in China.

The adoption of the Indigenous — or Chinese — style by the Republican government of Chiang Kaishek for its planned national reconstruction (*jianshe*) contributed to the consecration of this architectural form as China's national style. For much of the 1920s, Murphy served as an advisor to the Republican government in the planning of Guangzhou, Tianjin and Nanjing.[48] Along with some followers, he convinced state officials that Chinese cities, especially the capital of the new Republic, ought to retain what was "typically Chinese" by using a "purely Chinese architecture."[49] The transfer of the national capital from Beijing to Nanjing in 1928 heralded a period of intense building activity and presented a great occasion for architects to implement new architectural principles.

In 1930, the Chinese Architectural Research Society was founded, under the initiative of Liang Sicheng. Liang had returned from the University of Pennsylvania in 1928, having studied both classical Chinese architecture and Western urban planning. Throughout the 1930s, he conducted extensive surveys of historical buildings and greatly advanced knowledge of ancient Chinese design and construction techniques. Liang and his colleagues laid the foundations for the development of architectural research and education in China. Their research opened the door for a new era of modern Chinese architecture and allowed a new generation of architects to revive the spirit of the Confucian heritage in their search for an authentic Chinese form.

National reconstruction was interrupted by the Japanese invasion between 1937 and 1945, and the subsequent power struggles between the Kuomintang and the communists virtually stopped all architectural development in China for over a decade. Then, as Chiang Kaishek fled China to establish his provisionary seat in exile in 1949, he not only took China's entire gold reserve and most of the content of the National Museum, but he also appropriated the Indigenous style of architecture, bringing it to Taiwan where it continued to flourish as a national symbol.

Thus, overall, despite efforts at moving away from foreign influences, early attempts by Chinese architects at developing their own brand of modern architecture resulted in an essentially hybrid form, still largely influenced by foreign precedents and bearing the mark of the European training of its main proponents. Despite changes in designation and interpretation, the Indigenous style still retained many of the characteristics of the early Chinese style.

IDEOLOGICAL RECTITUDE AND STYLISTIC AMBIGUITIES: THE MAO YEARS

With the founding of the People's Republic in 1949, the new communist government was determined to refashion the image of socialist China, according to an ideological program that was both anti-Confucianist and pro-modern, which gradually became anti-foreign in spite of its dependence on foreign knowledge. Image-construction efforts were to be concentrated on the transformation of Beijing, which regained the status of national capital in 1948 after having lost it for twenty years. In a speech delivered in 1942 on the issue of artistic and literary works, Mao Zedong criticized the "slavish imitation of foreign fashion" and called for the need "to master our own national style," setting the pace for what was to become the slogan for artistic creation: "Socialist in Content, National in Style."[50] Throughout the Mao Years (1949–1976), the search for an architectural form fit for the new Communist nation was driven by sudden shifts in ideology and was filled with contradictions.

In the first years of communist power, architectural theory and practice fell into the hands of the key members of the Chinese Architectural Research Society, whose immediate architectural instincts were to promote the Indigenous style — in all its glory of glazed tiles and colorful decorations — as the symbol of the newly "liberated" nation. With this new interpretation, the style truly came to embody the constantly shifting course of China's twentieth century history: this neo-imperial architecture, which had first been developed to facilitate the infiltration of foreign missionaries and later became a symbol of Republican China, was now used to represent the glory of the socialist state.

Soviet advisors also contributed to the development of a Chinese socialist architecture by transplanting to China the national style they had established throughout the Soviet republics. This style consisted of a form of so-called "realist" monumental Neoclassicism, with decorative elements borrowed from local traditional architecture. It was said to promote a sense of national unity while expressing local identities. In China this new adaptation of the Indigenous style often translated into the crowning of Greco-Roman buildings by large, sweeping, tiled roofs — which were still held as strong symbols of "Chineseness." This new revolutionary style soon came to be known as the "Big Roof" style (*daping*). Liang Sicheng became an active proponent of the style, and his name became almost synonymous with the big roof. While acting as vice-chairman of the Beijing Urban Planning Committee in the 1950s, he is said to have turned down any building proposal that did not use big, traditional Chinese roofs.[51]

Early communist experimentation with the so-called Big Roof style was short lived, and was soon undermined by shifts in official ideology. After the launching of the First Five-Year Plan in 1953, which stated that economy and utility should be priorities in building construction, the revolutionary style came under attack for its impractical extravagances. In 1955, the official *People's Daily* published a series of articles exposing proponents of the style for having sacrificed functional requirements for visual concerns.[52] The same year the Chinese Ministry of Construction denounced the tendency in architectural practice to neglect the fundamental principles of economy and utility, while giving too much importance to formalism and aesthetics. Structures with large roofs and painted ornaments were condemned for being wasteful and bourgeois in principle. Liang Sicheng became the object of severe criticism for what was seen as his blunt disregard for the national austerity policy. Liang had neglected the economic, technical and practical aspects of architecture, and had stressed style rather than content, and the satisfaction of people's spiritual needs rather than their material ones. He became a common target for verbal attacks, as his intellectual point of view was denounced as capitalist idealism. After being labeled anti-party and anti-nation, Liang bowed, admitted his faults, and promised to comply with the party's policy.[53]

In light of such criticism campaigns, Chinese architects lived in fear of political reprisal, and did not dare venture outside of well-established political guidelines. Architectural creation became attributed to large design institutes rather than the effort of individual architects. At the occasion of the tenth anniversary of the foundation of the People's Republic in 1959, ten monumental buildings were constructed in Beijing, mixing elements from Soviet, modernist, and classical Chinese architecture, but without resorting to large sweeping roofs. One of these buildings was the Cultural Palace of Nationalities — a multifunctional structure reserved for the use of minority groups in Beijing. This building exemplifies early communist attempts at defining a Chinese national architecture with socialist characteristics. It was composed of a terraced structure dominated by a central tower in the Soviet realist style and topped by a series of small pavilion roofs of green glazed tiles **(Figure 7.4)**.

Also part of the "Big Projects for National Celebration" were the Great Hall of the People and the National Museum, which were erected on each side of the newly enlarged Tiananmen Square. These projects embodied some of the contradictions found in early attempts at representing the essence of the newly liberated China. Paradoxically, while their Neoclassical architecture resembled the temples of Western antiquity, their monumentality and orienta-

Figure 7.4. Cultural Palace of Nationalities, Beijing.

tion symbolically strengthened the city's ancient imperial axis, which recent urban renewal efforts had attempted to destroy **(Figure 7.5).**

After the departure of most Soviet experts in the early 1960s, Mao's Great Leap Forward led to a nationwide economic crisis followed by major cuts in public expenditures and a slowdown in construction throughout China. Until the end of the Mao era, architectural production was greatly reduced and mainly consisted of pragmatic designs deprived of any aesthetic pretense or reference to tradition. It was to take more than two decades for neotraditionalist approaches to architecture to reappear in China.

OPEN DOORS AND TILE ROOFS: POSTMODERNISM IN POST-MAO CHINA

After the initiation of Deng Xiaoping's Open Door policy in the late 1970s, the Chinese intellectual discourse was revived, with architectural circles leading the way in the debate over culture. Architectural production was no longer to be dictated by strictly ideological principles, nor practical requirements, but could now be regarded as something cultural and artistic.[54] Now relatively free to explore new directions, design professionals developed a variety of new approaches to counter the monotony of revolutionary architecture.

Figure 7.5. Great Hall of the People (1959).

During the late 1970s, increased contact with the outside world led to the emergence of a global architectural consciousness among Chinese professionals. The multiplication of international academic exchanges, the increasing mobility of Chinese design professionals, and the availability of foreign architectural magazines facilitated the permeation of international architectural trends in China. Early post-Mao architectural production was thus shaped by a desire to catch up with the great metropolises of the Asian and Western capitalist world and to project the image of a modern, advanced and prosperous nation. Chinese architects were mainly concerned with technological prowess in the creation of sleek, mirror-glass towers and postmodern extravaganzas. However, local technological limitations restrained the range of possibilities and compromised the quality of the buildings produced.

In the early post-Mao years, foreign architects, including Chinese living overseas, were brought back on the Chinese architectural scene by foreign investors to design international hotels and corporate headquarters. These architects often deplored the local obsession with Western-style architecture and tried to encourage the development of an architectural style more adapted to the local culture.[55] Despite their efforts, the international postmodern imagery remained predominant in Chinese architectural production for much of the 1980s. A 1987 *Beijing Daily* (*Beijing Ribao*) survey to identify the ten new buildings that best expressed the "spirit and material level of life in the 1980s" confirmed local aspirations for an image which would symbolize China's entry into the global village.[56] Among the ten finalists were several international hotels, such as the Great Wall Sheraton, Beijing's first joint-venture hotel designed by the American Welton Becket, and other postmodern look-alikes.

In the late 1980s, a great debate over China's "national character" and its relation to modernization emerged among Chinese intellectuals and soon spread to other segments of Chinese society. This "culture fever," as it was popularly referred to, was characterized by an intense reappraisal of China's cultural traditions and by discussions about the nature of modern Chinese identity.[57] As with the May Fourth Movement seventy years earlier, much of the debate revolved around the dilemma between tradition and modernity, Chinese and Western influences, and socialist and capitalist ideologies.

This cultural questioning coincided with the development of an historical conservation movement in China. Architectural historians and preservationists initiated a series of architectural surveys to assist policy-makers in implementing new regulations for the protection of cultural relics.[58] The preservation movement has also triggered renewed interest in developing a modern architectural style which will be able to preserve the character of the national capital.

Overseas Chinese, mainly Hong Kong and Taiwan residents who have in recent years returned to the mainland as investors or tourists, have brought another dimension of Chinese identity to the cultural debate of the 1980s, and they have influenced recent architectural production. Chinese in the diaspora were known for their loyalty to pre-Communist China and were often more attached to tradition than the people on the mainland.[59] Their contempt for the abuses of the Communist regime had led many of them to romanticize China's past from afar. Like their foreign counterparts, overseas Chinese architects also encouraged Chinese designers to resume the search for a modern architecture with Chinese characteristics, and they provided new models for mainland architects to emulate.

Taiwanese designers, who had for several decades continued the search for a modern Chinese architectural style, and whose work is now easily accessible in the mainland via Chinese-language architectural magazines, have recently provided a particularly important source of inspiration. Rather than using Chinese elements literally, these architects have reinterpreted traditional archetypes in a more abstract way, reminiscent of recent architectural developments in Japan.[60] Overseas Chinese architects working directly in the mainland also inspired a growing interest in local vernacular styles and regionalist borrowings. Chinese-born American architect I.M. Pei, after refusing to design a high-rise hotel in the center of old Beijing in the late 1970s, proposed what he saw as a "third way": an alternative to imitations of ancient Chinese architecture and to the Western glass box. His Fragrant Hill Hotel outside Beijing, built in the spirit of the vernacular courtyard houses of his ancestral Suzhou, provided a more spatial exploration of Chinese architectural precedents, and has had a lasting impact in terms of the development of regionalist architecture in China.[61]

One of the best-known followers of the regionalist approach in China is Prof. Wu Liangyong of Tsinghua University in Beijing, who in the late 1980s developed a series of housing prototypes based on the traditional multicourtyard model of southern China. Despite the popularity of his projects, widely published and well-known outside China, Wu's approach is often reminiscent of other attempts earlier this century to translate an essentialist vision of Chinese culture into architectural form. By blending historical preservation and the development of contextualist contemporary structures, Prof. Wu claims to "reconciliate the true nature of Chinese culture with the requirements of modern living" in order to create urban environments with a "truly Chinese sense of space."[62] Similarly, the work of regionalist architects such as I.M. Pei and Wu Liangyong is often based less on true contextualism than on a romanticization of regional prototypes — particularly the image of Suzhou, presented as the archetypal Chinese landscape — as they transpose this southern imagery into the northern environment of Beijing.

These early attempts at renewing the search for a Chinese architectural expression in the 1980s were thus shaped by a certain nostalgia and idealization of the Chinese past, which prevented them from going beyond formal and aesthetic considerations. In the early 1990s, efforts to revive the search for a modern architecture with Chinese characteristics were fueled by the Chinese government as part of a new strategy to win back political legitimacy and reinforce national unity.[63] Chen Xitong, a senior member of the Chinese Communist Party and mayor of Beijing from 1980 until 1995, is often identified as the main promoter of a postmodern brand of national style. For much of his time in office, it became virtually impossible to obtain a construction permit in Beijing without crowning a building with some "Chinese" decorative elements — generally a glazed tile roof **(Figure 7.6)**.

The main architectural consultant of the Beijing municipal government under Chen Xitong was Zhang Bo, chief architect of the Beijing Institute of Architectural Design and Research — one of the leading architectural firms of the capital city. As an old collaborator of Liang Sicheng and a proponent of the Big Roof style in the 1950s, Zhang strongly supported the reintroduction of elements of Chinese tradition in contemporary buildings. This new architecture, he claimed, would not only preserve Beijing's Chinese character under the threat of cultural homogenization posed by global capitalism, but it would ensure the city's modernization along internationally recognized lines.[64] Zhang thus worked closely with the state to enforce the aesthetic supremacy of this national style of architecture, and by the mid-1990s pavilion rooftops and Chinese ornaments were no longer limited to state institutions or national monuments but appeared indiscriminately on shopping centers, restaurants, public toilets, and telephone booths.

In 1994, a new survey on the state of contemporary architecture was initiated in Beijing, whose implicit goals were more about gaining legitimacy for the new national style than to gauge public opinion about architecture. The official *Beijing Daily* asked people to choose their favorite among a selection of 148 buildings in the national capital. Ten of the fifty winning buildings were the Great Monuments built in 1959 in the sinicized Soviet realist style; most of the remaining ones were recent Chinese revival buildings. Incidentally, senior architect Zhang Bo had been directly involved in the design of eleven of the fifty winning building entries, including the Palace of Nationalities, which ranked first on the list. The biased nature of the survey was made explicit in the list of preselected candidates, none of which were built between 1960 and 1980, and only three of which did not exhibit traditional features.

Figure 7.6. The Beijing West Station.

Clearly, the new official architectural agenda sought the reversal of the practices of the Mao era for a more regionalist approach. In a publication the organizing committee stated that rows of similar box-like structures were no longer desirable in Beijing, and buildings exhibiting national and regional features should be used to ensure the modernization of the Chinese capital in the least disruptive manner.[65]

In recent years a new generation of high-rise buildings exhibiting a whole range of variations on the pagoda theme have appeared throughout Beijing. Such architectural developments have been widely criticized, both in China and abroad, for not only failing to preserve the character of the old capital but also for their trivialization of Chinese tradition. Their stereotypical use of traditional architectural features, reduced to mere icons of "Chineseness," have given rise to allegations about the Chinatownization of Beijing, Disney-style.[66] Beijing residents have themselves derided new buildings as "wearing a Western suit with a Chinese hat."

Until recently such a style continued to be sponsored by the state in its attempt to revamp the image of Beijing as China's national capital. Although Chinese authorities may have first sought its political value in promoting unity in the face of rising separatist movements among China's ethnic minorities, they have also embraced the style for its potential economic value. Chinese authorities have made great efforts to court foreign investors and bring tourists back to China after the lean years following the 1989 Tiananmen massacre. The hybrid image produced by the new architecture may thus play a double role in combining traditional elements to attract visitors in quest for exoticism with a modern image of prosperity to reassure potential investors. And like Chinese merchants in North American Chinatowns, city authorities now capitalize on the world idea of "Chinese" for monetary gains.

CONCLUSION

Today the Chinese style of architecture has become fully incorporated into the pan-Chinese collective consciousness, to the point of amnesia about its origins. Indeed, contemporary Chinese scholarship on China's modern architectural history either downplays or entirely ignores the role of foreign architects in developing the Chinese national style. The origins of the style are generally traced back to Lu Yanzhi's 1925 design for the Sun Yatsen Mausoleum in Nanjing. Although a few publications actually mention Murphy's contribution to Chinese architecture, they generally invert the roles by suggesting that his adaptation of Chinese traditional architecture to modern requirements was inspired by the work of Lu and other of his apprentices. Today any admission that foreign architects actually invented the Chinese national style is simply not acceptable.[67]

The historical development of this architectural style, from its origins as an Orientalist import to its adoption as a national symbol by Chinese across the ideological spectrum, has taken the allure of an epic story, a metaphor for the search for a modern Chinese identity. Filled with contradictions, challenges and reevaluations, it illustrates the constructed and often instrumental nature of collective identities. Throughout this century different actors, both within and without Chinese society, have tried to manipulate collective representation and simplify the complexity of a whole society to serve their ideological and economic interests. And despite continuous efforts, they have been faced with the impossibility of finding a unique material expression which could represent the Chinese as a whole. This enduring search for an architectural expression which could embody the character of an entire people has rested on an essentialist notion of the nature of collective identity as monolithic, definable, and capable of being translated into built form. It has also rested on a blind faith in the representational power of architecture and its capacity to communicate simple, unequivocal meanings.

This century-long search for a Chinese architectural expression clearly demonstrates that the constitution of collective identity is far from being a self-contained event, but can be truly global in nature. Different attempts at representing Chinese identity have been shaped by ideas, agents and events coming from across the temporal and spatial spectrum, constantly reformulated over the years to serve new needs and purposes. Many external factors have been internalized, becoming part of the local reality and adding new dimensions to that reality. With China's opening to the world, collective identity definition is more than ever the product of the interaction of diverse competing forces.

Recent architectural developments in Beijing thus only represent the culmination of one hundred years in the pursuit of a Chinese architectural identity. Today the original dilemma between East and West, past and present, local and foreign, remains. Maybe it is this constant questioning, this continuous reevaluation, and this inherent contradiction that is the essence of Chinese modernity, perfectly embodied in this elusive architectural style.

NOTES

All photographs are from the author's personal collection. The author acknowledges the generosity of *Tien Hsia* and *The China Academy of Building Research*.

1. E.N. Bacon, *Design of Cities*, New York, Viking Press, 1976, p.233; and S.E. Rasmussen, *Towns and Buildings Described in Drawings and Words*, Liverpool, University Press of Liverpool, 1951, p.i.

2. Chinese historian Frederic Wakeman once recalled how upon a visit to the newly restored Confucius temple in Beijing, an eminent Chinese scholar in his company disputed the site's superficial, tourist-oriented preservation, by saying, "This is no longer Beijing. It is Chinatown."

3. L.C. Arlington and W. Lewisohn, *In Search of Old Peking*, Peking, Henry Vetch, 1935, p.v.

4. J. Bredon, *Peking: A Historical and Intimate Description of Its Chief Places of Interest*, Shanghai, Kelly and Walsh, 1919, p.2.

5. For a discussion of issues of identity, see D.B. Massey, *Space, Place, and Gender*, Minneapolis, University of Minnesota Press, 1994; S. Hall and P. Du Gay, *Questions of Cultural Identity*, London, Sage, 1996; M. Castells, *The Power of Identity*, Malden, MA, Blackwell, 1997. A.R. Pred and M. Watts, *Reworking Modernity: Capitalisms and Symbolic Discontent*, New Brunswick, NJ, Rutgers University Press, 1992; and A.R. Pred, "Out of Bounds and Undisciplined: Social Inquiry and the Current Moment of Danger," *Social Research*, vol.62, no.4, 1995, pp.1065–1092.

6. See R.J.C. Young, *Colonial Desire: Hybridity in Theory, Culture and Race*, London, Routledge, 1995; M. Featherstone, *Global Culture: Nationalism, Globalization, and Modernity: A Theory, Culture & Society Special Issue*, London, Sage Publications, 1990; and A. Appadurai, *Modernity at Large: Cultural Dimensions of Globalization*, Minneapolis, University of Minnesota Press, 1996.

7. Young, *Colonial Desire*, p.27. H.K. Bhabha, *The Location of Culture*, London, Routledge, 1994; and "Life at the Border: Hybrid Identities of the Present," *New Perspectives Quarterly*, vol.14, no.1, 1997, pp.30–31.

8. Occasionally, Chinese master builders erected modern structures to serve the needs of foreign merchants and entrepreneurs in the coastal trading ports, but racial prejudice had often prevented them from participating in the building trade. For example, in 1916 a consortium of foreign architects pressured the Shanghai Municipal Council to impose strict regulations which required that all builders of modern structures hold official registration as professional engineers or architects, thus excluding most Chinese builders. See C.H. Godfrey and H. Ross, *Some Notes on the Shanghai Building Rules, Proceedings of the Engineer Society of China and the Report of the Council*, vol.16, 1916–1917, as cited in J.W. Cody, "Henry K. Murphy: An American Architect in China, 1914–1935," Ph.D. diss. in Historic Preservation Planning, Cornell University, 1989, pp.96–97.

9. The term "comprador" generally referred to a Chinese agent engaged by a foreign establishment to act as an intermediary in business affairs. "Compra-doric," therefore, alluded to the use of Neoclassical architecture (with elements such as doric columns) by foreign builders in China.

10. H. Hussey, *My Pleasures and Palaces: An Informal Memoir of Forty Years in Modern China*, Garden City, NY, Doubleday, 1968; and G. King, "Utilisation of Chinese Architectural Design in Modern Buildings: The Rockefeller Foundation's Hospital Plant at Peking," *The Far Eastern Review*, August 1919, pp.562–563.

11. "Dedication Ceremonies and Medical Conferences," *Peking Union Medical College: Addresses and Papers*, 1916, p.57.

12. Rockefeller Foundation, *Annual Report*, 1917, p.224.

13. M. Bullock, *An American Transplant: The Rockefeller Foundation and Peking Union Medical College*, Berkeley, University of California Press, 1980; and *Zhongguo jianzhushi (History of Chinese Architecture)*, Beijing, Zhongguo jianzhu gongye chubanshe, 1986.

14. H.K. Murphy, "An Architectural Renaissance in China: The Utilization in Modern Buildings of the Great Styles of the Past," *Asia and the Americas*, June 28, 1928, pp.468–475. For material on Murphy, I am greatly indebted to Jeffrey Cody, whose impressive dissertation on the life of this American architect has been central to my understanding of his role in architectural developments in China. See J.W. Cody, "Henry K. Murphy."

15. Ibid., J.W. Cody, pp.88–98.

16. Ibid., pp.40–41.

17. On social construction at the world's fairs, see T. Mitchell, *Colonising Egypt*, New York, Cambridge University Press, 1988; B. Benedict, *The Anthropology of World's Fairs: San Francisco Panama Pacific International Exhibition of 1915*, Berkeley, Scolar Press, 1983; and R.W. Rydell, *All the World's a Fair: Visions of Empire at American International Exhibitions, 1876–1916*, Chicago, University of Chicago Press, 1984.

18. For example, at the San Francisco's Panama Pacific International Exposition of 1915, a concession called "Underground Chinatown" run by Sid Grauman, a local entrepreneur, depicted a "chamber of horror" with Chinese in an opium den. See Benedict, *The Anthropology of World's Fairs*, p.51; and Rydell, *All the World's a Fair*, p.228.

19. "The Centennial Exposition," *Public Record*, Philadelphia, October 5, 1876, p.4.

20. A. Quantin, *L'Exposition du Siècle*, Paris, Le Monde Moderne, 1900, p.193–194.

21. Murphy, "An Architectural Renaissance," p.470.

22. Ibid., p.468.

23. "Yale-in-China: Western University Movements in the New Republic," *The Far Eastern Economic Review*, July 14, p.81.

24. Ibid., p.82.

25. Cody, "Henry K. Murphy," p.26–48; and Murphy, "An Architectural Renaissance," p.468–475.

26. Cody, "Henry K. Murphy," p.148-149.

27. Students of Orientalism also pointed out the necessity for the West to construct an "Other" as primitive, exotic, and thus inferior to reinforce confidence in European identity as superior. See E.W. Said, *Orientalism*, New York, Pantheon Books, 1978.

28. Murphy & Dana Architects, "Tsing Hua University Peking China," *The Architectural Review*, vol.4, no.10, 1916, pp.180–182; Cody, "Henry K. Murphy," p.128.

29. E.C. Jones, "Fukien Christian University: Transforming the Mind of China from an Educational Base in Foochow," *The Far Eastern Review*, May 1920, pp.234–236. Similarly, in 1919, the newly appointed president of Yanjing University in Beijing, John Leighton Stuart, requested that the evangelical nature of the university be invisible to the Chinese people, and thus demanded that it be built in the Chinese style. See Jin Feng, "Orientalism and the Twentieth Century Chinese Architectural Dilemma," paper presented at the 5th conference of the International Association for the Study of Traditional Environments (IASTE), Berkeley, 1996.

30. Jean Dethier developed the term "psychic pacification" in his study of colonial Morocco to explain how colonialists attempted to gain control over indigenous popula-

tions through the fostering of an excessive fascination for their past, which would petrify identities and reduce populations to a nonintellectualized expression of their own culture. See J. Dethier, "Evolution of the Concepts of Housing, Urbanism, and Country Planning in a Developing Country: Morocco 1900–1972," in L.C. Brown (ed.), *From Medina to Metropolis*, Princeton, NJ, Darwin Press, 1973.

31. Dethier, "Evolution of the Concepts of Housing"; A.D. King, *Colonial Urban Development: Culture, Social Power, and Environment*, London and Boston, Routledge & Paul, 1976; and G. Wright, *The Politics of Design in French Colonial Urbanism*, Chicago: University of Chicago Press, 1991.

32. W. Kaplan, "Regionalism in American Architecture," in E. Cumming and W. Kaplan (eds.), *The Arts and Crafts Movement*, New York, Thames and Hudson, p.107.

33. For different accounts of the development of Chinatown architecture, see C.L. Yip, "San Francisco's Chinatown: An Architectural and Urban History," Ph.D. diss., University of California, Berkeley, 1985; "Associations, Residence and Shop: An Appropriation of Commercial Blocks in North American Chinatowns," in E.C. Cromley and C.L. Hudgins (eds.), *Gender, Class and Shelter: Perspectives in Vernacular Architecture*, Knoxville, University of Tennessee Press, 1995; and D.C. Lai, "The Visual Character of Chinatowns," *Places*, vol.7, no.1, 1990, p.29.

34. Yip, "San Francisco's Chinatown," p.173.

35. Ibid., p.197.

36. Lai, "The Visual Character of Chinatowns," p.29.

37. After the anti-foreign 1900 Boxer Rebellion, during which several foreign legions within China were attacked, the Chinese government was asked to repay an indemnity to the foreign countries that had suffered losses. The United States then decreed that part of the Boxer indemnity remitted in 1908 should be paid in the form of scholarships for Chinese students to study in the U.S. See "Returned Chinese Students: Filling High Positions in China," *The Far Eastern Review*, November 1917, pp.740–741.

38. At the time, knowledge of Chinese architecture in the West was very limited, as few historians deemed it worthy of much attention. Architectural historian Sir James Fergusson, in his three-volume history of world architecture, presents the Chinese as "singularly deficient in every element of greatness, either political or artistic." Their architecture, he adds, "is totally wanting in dignity and grandeur of conception" (Fergusson, *History of Indian and Eastern Architecture*, London, John Murray, 1876, pp.688–710). Sir Banister Fletcher similarly contended that "the Chinese had very little feeling for architecture," and that Chinese architecture was "peculiar," "non-historical," and had been "practically stationary for many centuries" (Fletcher, *A History of Architecture: On the Comparative Method of the Student, Craftsman and Amateur*, 5th ed., New York, Scribner's Sons, 1905, pp.808–809).

39. "Returned Chinese Students: Filling High Positions in China," pp.740–41.

40. Scott, *Literature and the Arts*, p.112.

41. Cody, "Henry K. Murphy," pp.208–209.

42. Jin, "Orientalism."

43. W.H. Chaund, "Architectural Effort and Chinese Nationalism: Being a Radical Interpretation of Modern Architecture as a Potent Factor in Civilization," *The Far Eastern Review*, vol.9, no.8, 1919, p.533.

44. Chaund, "Architectural Effort," p.534.

45. Cody, "Henry K. Murphy," p.334.

46. See D. Doon, "Architecture Chronicle," *T'ien Hsia Monthly*, vol.3, November 1936, p.360; G.D. Su, *Chinese Architecture, Past and Contemporary*, Hong Kong, The Sin

Pih Amalgamated, 1964, p.136; and Y.C. Lu, "Memorials to Dr. Sun Yat-Sen in Nanking and Canton," *The Far Eastern Review*, vol.25, no.3, 1929, pp.97–101.

47. D.B. Stewart, *The Making of a Modern Japanese Architecture: 1868 to the Present*, Tokyo and New York, Kodansha International, 1987; and Cody, "Henry K. Murphy," p.27.

48. "Canton Adopts New 'City Plan': Chinese Aspects Retained in Layout Designed by a New York Architect," *New York Times*, March 13, 1927, p.13; "American Chosen to Advise Nanking," *New York Times*, October 15, 1928, p.25.

49. J.W. Cody, "American Planning in Republican China, 1911–1937," *Planning Perspectives*, vol.11, 1996, pp.339–377.

50. Quoted in M.S. Samuels and C. Samuels, "Beijing and the Power of Place in Modern China," in J.A. Agnew and J.S. Duncan (eds.), *The Power of Place: Bringing Together Geographical and Sociological Imaginations*, Boston, Unwin Hyman, 1989, pp.214–221.

51. J. Zha, "Beijing: A City Without Walls," *Antioch Review*, vol.52, no.3, 1994, pp.390–416.

52. In February 1955, the *People's Daily (Renmin Ribao)* published a caricature deriding recent architectural tendencies, where the defunct Empress Dowager was shown patting a present-day architect on the back, saying, "You surely know how to spend the money: I never thought of using glazed tiles for the kitchen when I built the Summer Palace." Scott, *Literature and the Arts*, p.117. See also the *People's Daily* editorials of March 28 and June 19, 1955.

53. Su, *Chinese Architecture*, pp.158–165.

54. Zha, "Beijing," p.410.

55. For example, see S. Stuebing, "Building in China: The Promise, The Problems," *Progressive Architecture*, March 1986, pp.31-33; and D. Cohn, "China: The Search for National Forms and Modern Techniques," *Architecture: The AIA Journal*, September 1985, pp.78–81.

56. M. Hood, "China Still Suffers Identity Crisis as It Tries to Establish Its Architecture," *South China Morning Post*, December 6, 1987, p.17. Despite the newspaper deputy editor's claim that this was to be a democratic selection process, a panel of experts with veto power was nonetheless selected in case the masses made an "inappropriate" choice.

57. W. Jing, *High Culture Fever: Politics, Aesthetics and Ideology in Deng's China*, Berkeley, University of California Press, 1996.

58. One of the strongest advocates of this movement has been Liang Conjie, the son of Liang Sicheng, who has fought without much success for the preservation of old Beijing, which he sees as threatened by recent modernization.

59. W. Ling-Chi, "Roots and Changing Identity of the Chinese in the United States," *Daedalus*, vol.120, no.2, 1991; and W.-M. Tu, *The Living Tree: The Changing Meaning of Being Chinese Today*, Cambridge, MA, American Academy of Arts and Sciences, 1991.

60. M. Pawley, "Profile: Seeking the Architecture of New China: C.Y. Lee and Partners," *World Architecture*, vol. 54, March 1997, pp.78–117.

61. M. Gu, "A Chinese View of Fragrant Hill," *Architecture*, November 1983, pp.88–89.

62. L. Wu, "A Probe into the Organic Renewal of Urban Cells and the New Type Courtyard House," *China City Planning Review*, June 1990, pp.3–23; and L. Wu, "After a New Prototype: Rehabilitation of Residential Area in Old Beijing," *JIIA*, March 1991, pp.31–38.

63. See C. Feng, "Order and Stability in Social Transition: Neoconservative Political Thought in Post-1989 China," *China Quarterly* 1997.

64. J. Lin, "Architects with Designs on Tradition," *China Daily*, June 28, 1994, p.5.

65. Architectural Art Committee of the Capital, *New Architectures in the Capital: New Architectures with National Styles and Well-Received by the Masses*, Beijing, Beijing Publishing House, 1995, p.257; "Vote for Buildings," *China Daily*, March 10, 1994, p.3; "Classical Buildings," *China Daily*, April 29, 1994, p.3; and J. Lin, "Securing the Look of Old Beijing," *China Daily*, May 31, 1994, p.11.

66. J. Rykwert, "Between Mao and Disneyland," *Times Literary Supplement*, September 10, 1993, pp.16–17; "Provided They Have Curly Tops," *The Economist*, July 16, 1996, p.80.

67. Cody, "Henry K. Murphy," p.327; and Jin, "Orientalism."

8

Building Culture in Divided Berlin: Globalization and the Cold War

Greg Castillo

Globalization, often seen as a herald of the twenty-first century, was given a dress rehearsal during the 1950s in a tug-of-war that tore global culture in two. The Cold War's victories and defeats are conventionally scored in terms of nuclear warheads, campaigns to dominate the Third World and outer space, and the fluctuating diplomacies of belligerence and detente. But culture — including that of environmental design — was also conscripted by the United States and the Soviet Union in their battle for preeminence. Each superpower promoted its version of a modern, industrialized society within its own hemisphere of influence, and attempted to do so in its adversary's as well. Culturally specific definitions of progress, tradition, national identity, and international community were not only exported as ideas, but also as ideal settings for postwar life, from cities and streets to homes and their furnishings. These contending material cultures were propagated to establish the characteristic subjectivities of two international orders. One was based on Stalin's vision of "smokestack socialism." Its party-managed agenda emphasized heavy industry and the creation of proletarian subjects attuned to the needs of a Soviet-style command economy. The "other" was an alliance of liberal capitalist democracies, linked by open markets and unrestricted trade, proposed by Woodrow Wilson and instituted after World War II by the U.S. Marshall Plan. Today's processes of globalization and cultural hybridity were foreshadowed in the expansion of these two competing hegemonic orders, as well as in the tactics of accommodation or resistance enacted by those on the receiving end. Without an assessment of how these alternative visions of twentieth century civilization were transmitted and experienced, our understanding of globalization's broader history remains incomplete.

Germany's postwar reconstruction provided a perfect laboratory for the deployment of environmental design as a tool for social reform — and, simultaneously, as a weapon in the Cold War's ideological arsenal. The victors' mandate was to regenerate a vanquished nation from the ground up. Institutions and traditions associated with the discredited Nazi past were to be obliterated. For postwar planners from the occupying powers, this would amount to nothing less than a reinvention of the German "national character." Open resistance to their programs of reform was unlikely; it easily could easily be branded as unreconstructed Nazism. With the war's survivors prostrate and civil society all but extinguished, conditions were in place for a radical transformation of German social structure and its built artifacts.

After a brief flirtation with postwar cooperation, joint allied military occupation gave way in 1949 to the creation of two separate German states: one affiliated with the U.S., the other with the U.S.S.R. Divergences in environmental-design practices between these two new Germanies now not only expressed a break with the Nazi past, but also the assimilation of a template for modern life prescribed by a patron superpower. The Zhdanovshchina, a cultural revolution of Russian provenance, transformed East Germany's architectural establishment in 1950. Soviet advisors equated aesthetic modernism with American imperialism and the intelligentsia's alienation from proletarian class consciousness. The prescribed antidote was Socialist Realism: a unified formulation of the arts developed in the 1930s in the U.S.S.R., which advocated an architecture "socialist in content and national in form."[1] In West Germany, American advisors provided evidence for Soviet propaganda by actively promoting the revival of Bauhaus modernism. It was framed as the native dialect of a broader "International Style" which served as a shared symbol of progress among nations in America's "informal" postwar empire.[2] As different as the building cultures of the two Germanies were in style, they shared methodological similarities in their constitution of a contemporary design heritage. Specific local design traditions, said to represent "authentic" national traits unpolluted by Nazism, were merged with cultural norms imported from a superpower mentor. The result was a pair of antithetical models for postwar German urbanism and urban subjectivity. Both of these opposing design epistemologies were supported by ties to a scripted past which could be used to defuse charges of cultural imperialism leveled by the other side.

There is no better place to examine the Cold War's processes of cultural globalization and hybridity than in Berlin. By 1948, strategies of ideological containment had turned Europe into an unlikely set of nesting boxes: a divided continent, which encased a divided Germany, which contained a divided metropolis. Traversed by the fault line of geopolitical confrontation, Berlin assumed a variety of singular Cold War roles, including its use as a showcase for the West's International Style and its Socialist Realist "other." The permeable border crossing between capitalist and socialist versions of the same city — a condition that changed only with the construction of the Berlin Wall in 1961 — provided both camps with what one U.S. Department of State official called "unlimited propaganda possibilities."[3] Yet the Cold War's battle of styles, even during its heyday, was not apparent in the overall rebuilding of East and West Berlin. During this period of economic resuscitation, patch-up repairs and bare-bones utility ruled reconstruction. Architectural form served a declamatory function only in projects invested with particular representational significance. A pair of extravagant housing developments in East and West Berlin, the Stalinallee and the Hansaviertel, provide precise illustrations of the divergence of postwar German architecture and urbanism. West Berlin's Hansaviertel was built specifically as a housing exhibition; the Stalinallee was one by default, given the blurring of private and public spheres in East Germany as all aspects of society were subordinated to the economic and ideological mandates of Stalinism. At divided Berlin's two residential showplaces, fundamental axioms about modernity's socialist and capitalist guises were put into practice and "lived out." Appropriately enough, these competing model cities were built on opposite ends of an imaginary thoroughfare, Berlin's planned east/west axis, which had eluded attempts at realization for half a century. Although sited along an historical urban alignment and set less than four miles apart, the Stalinallee and Hansaviertel remained separated by two city limits, a national border, and the divide between two postwar hemispheres.

TWO CITIES OF TOMORROW

The Stalinallee was the building of East German socialism made tangible. It made an indelible impression, rising, in the words of an eyewitness, like "a sculpture, isolated in the midst of a cityscape which had just barely been cleared of the wreckage of war."[4] Begun in 1952 and largely completed by 1957, the Stalinallee's linear urbanism framed a boulevard conceived in the grand European tradition. In keeping with Moscow precedent, the avenue was designed to serve two modes of circulation vital to the function of a socialist capital. Six lanes of pavement provided daily access for transportation in and out of the city center. On political holidays, the street was converted from a vehicular artery to one that carried ideological traffic. As a processional route for socialist celebrations, the Stalinallee substantiated East Berlin's claim as the capital of German socialism with a display of mass spectacle **(Figure 8.1)**.

Plans for the Stalinallee's individual structures emerged from six state-funded design groups, each organized under a senior partner: Egon Hartmann, Hermann Henselmann, Hans Hopp, Kurt Leucht, Richard Paulick and Karl Souradny. This collective organization of architectural practice and its standardized approach to the design of the street's constituent elements were transmitted from Moscow to Berlin through a 1950 Soviet retraining program for key members of East Germany's Ministry of Construction. Tenets of Soviet urban design were demonstrated for the first time in Germany by the Stalinallee's harmonic ensemble of superblock facades. Socialism's bequest to architects and planners lay in the ability to realize a visionary concept of this radical reconfiguration of Berlin's prewar pattern of city blocks and building lots: powers to condemn and expropriate now were unhindered by the constraints of private ownership.

At the Stalinallee, design idioms laced with regional allusion were the Socialist Realist nostrum for what was decried as modernism's homogenization of national

Figure 8.1. Poster commemorating the Month of German-Soviet Friendship, 1952. Under the twin towers of the Strausbergerplatz (architect: Hermann Henselmann, 1952-1954), representatives of various aspects of the economy — but of only two age cohorts of the population, that is, young adults and young children, symbolizing production and reproduction — are depicted marching down the Stalinallee in a celebratory procession. (Courtesy of Deutsches Historisches Museum, P 96/434.)

cultures. Neoclassical detail, rendered in a rich tan travertine, was said to reflect a critical appropriation of the work of Prussia's most revered architect, Karl Friedrich Schinkel. Above a plinth dedicated to socialist commerce rose six-to-seven stories of flats sheathed in cream-colored tile made at the Meissen works — an establishment once patronized by European nobles for its exquisite porcelains, now transformed into a manufacturer of building materials for a proletarian clientele. The Stalinallee's ceramic cladding was enlivened with texts and images which celebrated the building of socialist Germany. Construction workers rendered in bas-relief invoked the heroism of reconstruction; shields of the Five-Year Plan announced a planned economy; and stylized rosettes and sheaves of grain (inspired by folk art but inflated here to architectural scale) epitomized Socialist Realism's proclaimed fusion of all "progressive" modes of expression into a single creative method. For exponents of the Stalinist aesthetic, these buildings were textbook examples of national tradition infused with socialist content. But to critics, the question was *which* national tradition — Prussian or Russian? The Stalinallee's stockpile of Classical components, shared internationally for centuries, eluded a definitive pedigree.

Western detractors compared these "workers' palaces" and their ornate street frontages with Berlin's notorious *mietskasernen* (literally, "rental barracks"):

a nineteenth-century housing type that used facades of bourgeois decorum to disguise miserable workers' flats packed into built-out urban blocks punctured with meager light wells. Accusations of such unseemly ancestry, of great propaganda value to the West, were ultimately proven false when the Stalinallee was viewed from the air **(Figure 8.2)**. The street's masonry monoliths turned out to be more like inhabited false fronts, stretching a superblock in length but only two rooms deep. These slender housing slabs served two distinct aspects of urban reform. Like their counterparts in the German social housing movement of the 1920s, they maximized each dwelling unit's exposure to fresh air and sunshine. But the building format also advanced Socialist Realism's aesthetic reform of postwar Berlin. From a vantage point bracketed by a gleaming progression of tall Neoclassical backdrops, one saw the new capital city of German socialism not as it was, but as it could be. Within this fragment of utopia, the curtain of Stalinallee facades screened from view Berlin's bleak patchwork of inhabited ruins. Visitors were told: "Multiply this image by one thousand, multiply by one thousand the generous appointments of the street, the mechanical comforts of the housing, the

Figure 8.2. Aerial view of the Stalinallee rising amid the ruins of East Berlin's Friedrichshain district in 1954. The oval-shaped Strausbergerplatz is at the left. One of Henselmann's twin towers remains encased in dark construction scaffolding. The columned facade at the far right, slightly obscured by landing gear, is that of the Deutsche Sportshalle (architect: Richard Paulick, 1951). (Courtesy of Berlin Landesbildstelle, 24 LuA 33 797.)

attractive shops, social facilities and restaurants, and you get a general idea of the good life in the socialist residential district of the future."[5] Yet, although conforming to the representational tenets of Socialist Realist literature, this approach to urban reconstruction made for poor public-housing policy. To build the proletarian luxury dwellings of the Stalinallee, the East German government in 1952 allotted an average per-unit cost four times higher than the target for the new state-built apartments it had endorsed just two years earlier.[6]

From architectural symbolism to modes of construction and consumption, the Stalinallee choreographed a transfer of culture and knowledge from Russia to its German ally. The street's plan was fine-tuned in 1951 with advice from the city architect of Moscow, Aleksandr Vlasov, and the vice-president of the Soviet Academy of Architects, Sergei Chernyschev. At the cornerstone-laying ceremony, East German Minister President Otto Grotewohl announced the opening of a route leading from East Berlin's Alexanderplatz to Moscow, the hub of world socialism. One of the first housing blocks built was dedicated to the Society for German-Soviet Friendship. A Stalin Prize-winning mason named Korolev, whose eponymous high-speed bricklaying technique was hailed as a breakthrough in socialist productivity, visited from Russia to set this building's last thousand bricks in a widely publicized construction spectacle. Strolling along the finished boulevard, East Berliners could feel themselves part of a great postwar world that spread out before them in one direction — east. After refreshments at the Budapest Restaurant or Cafe Warsaw, window shoppers could peruse barrels of Bulgarian feta cheese and bottles of Tapolka and Kiravi wines, the luxury imports of an autarkic Soviet-bloc economy. Just down the street, sports events held at the new Deutsche Sportshalle brought together East German and other Eastern European athletes with North Koreans, Red Chinese, and the peoples of the U.S.S.R.'s "family of nations." From a pedestal set back from the boulevard, a statue of Stalin presided over these cultural and economic intersections of a socialist hemisphere that depicted itself as a self-sufficient world.

This synecdochic geography, in which a hemispheric fragment substituted for a global realm, was repeated in West Berlin's model-city project: the Hansaviertel. Its 1,256 dwellings were the product of Interbau, the 1957 International Building Exposition. The project began in 1953 with a design competition for the reconstruction of a heavily bombed nineteenth century residential district on the western edge of Berlin's "central park," the Tiergarten. A clarification of the competition's goals by a member of Berlin's city senate specified what was desired: "a lucid declaration of the architecture of the Western world. It should demonstrate what we understand to be modern urbanism and proper housing, in contrast to the false ostentation of the 'Stalinallee'."[7]

For a journalist writing in *Die Neue Zeitung* ("the American newspaper for the German people"), West Berlin's new calling as a "City of International Architecture" implied an invitation to the world's best architects not only to create a model neighborhood, but also to found a new design school as well. The motto "All of Berlin an enormous Dessau Bauhaus!" would identify both projects.[8] Although the school never materialized, the global agenda of the modernist exposition did (**Figure 8.3**). Interbau provided building commissions for such luminaries as Alvar Aalto, Le Corbusier, Walter Gropius, Arne Jakobsen, Oscar Niemeyer and Hans Scharoun; and it assembled architectural talents from Austria, Brazil, Denmark, England, Finland, France, Israel, Italy, Sweden, Switzerland, the United States and West Germany. It also included the work of a little-known local designer, Brigitte D'Ortschy, who had spent a semester in North Carolina a few years earlier learning American design practices in a retraining program sponsored by the U.S. High Commissioner for Germany at precisely the moment that East German architects were being reoriented in Moscow to the norms of Socialist Realism. Interbau's participating architects represented a geographic entity known as "the free world," and were called upon to demonstrate "the free world's technology and creative strength in its wide variety of forms."[9] Their work in "free Berlin" described a setting within that imaginary geography — the city's awkward physical location, embedded within the East German republic, notwithstanding.

Interbau was conceived as a concrete reproach to the Stalinallee's Neoclassicism. "One thing is certain," declared West German President Theodor Heuss about the exposition, with his glance cast eastward: "any yammering about 'tradition' will receive no response."[10] The design team responsi-

Figure 8.3. Aerial view of the Interbau reconstruction of Berlin's Hansaviertel district in 1962. One edge of the development is bounded by the elevated train tracks, seen as a dark arc extending from the lower left to the upper right side of the photo. The site's other edges abut Berlin's Tiergarten. (Courtesy of Berlin Landesbildstelle, 4 LuA 83 467.)

ble for the Hansaviertel site plan, Willy Kreuer and Gerhard Jobst, proposed that individual structures be disposed "in open nature . . . and, through this unforced arrangement, should prove an obvious antipode to buildings designed for dictatorial needs."[11] "Comparing the site plan and sketches with the Stalinallee ensemble," an authority at East Berlin's Bauakademie shot back, "no one of sound judgement could possibly find this fantasy from an extinct world beautiful."[12] Interbau's invited architects worked within the parameters of a general plan which determined the site and size of buildings; otherwise, designers had free reign. Housing was built in rows, slabs and towers with facades composed in themes and variations on the structural grid. Walter Gropius, for one, tried to build consensus with the designers of a group of neighboring structures for a unifying palette of exterior materials, but to no avail.[13] The result was a random-looking assortment of buildings scattered freely amid park greenery. "Plenty of primadonnas looking for a conductor!" was the verdict of the Italian architect Bruno Zevi, writing for *L'Espresso*.[14] Visual incoherence was the price paid to create an urban metaphor for Western individuality and freedom.

The Hansaviertel's symbolic duties included the representation of a technologically advanced consumer economy. As a laboratory for new construction, Interbau incorporated a wide spectrum of innovative building methods — prefabricated panels, gas-injected concrete, and a variety of synthetic insulating and waterproofing materials, just to name a few. In the realm of domestic engineering, "what just a few years earlier would have been described as luxuries . . . now should be taken for granted," the exhibition catalog read. Bathrooms were appraised on aesthetic as well as hygienic grounds; kitchens were fully outfitted and, for the most part, all-electric; and even refrigerators were considered standard equipment — far from the norm for West Berlin homes in the mid-1950s. Residential interiors displayed a variety and adaptability of space planning unknown in the East. Apartments ranged from efficiency flats to two-story maisonettes. Open floor-plans were tempered with movable panels which captured shifting configurations of interior space for changing uses. Transcending the specifics of individual designs, what all Interbau residences shared was an increased solicitude for leisure, privacy, and technological indulgence — hallmarks of the comfortable life style made possible by the *Wirtschaftswunder*, West Germany's "economic miracle." At the reconstructed Hansaviertel, modernism was domesticated and harnessed to consumer desire to depict the future of West German housing and the triumph of the Marshall Plan.

Interbau's traffic planning protested the Stalinallee's panoramic, all-purpose streetscape and recapitulated the ideals of modernist urbanism. Housing units faced away from the street whenever possible. Pedestrian paths and a separate road network dedicated to motorized traffic linked residences to the community. Although the Stalinallee glorified the proletariat with a boulevard of aristocratic grandeur, the primacy of the automobile on Interbau's street system trumpeted the emergence of an all-encompassing middle class. A municipal publication released in conjunction with the exhibition linked private transportation to a trajectory of social progress as reductive as anything ever generated by Marxism-for-the-masses. "The worker has become a petit-bourgeois [*Kleinbürger*], the haute-bourgeois [*Großbürger*] long ago became a citizen [*Bürger*] — a society of the classless middle is in the making." The only truly enfranchised citizen of West Berlin's City of Tomorrow, however, was to be the bearer of a driver's license, zooming along "new, much wider, much straighter, much longer streets." "And the pedestrian? The new street can't deal with incorrigible Neanderthals. Whoever has a destination should sit in a car; whoever has none is a promenader and immediately heads to the nearest park."[15] Interbau's headlong rush into an automotive future was rash for West Berlin: at the time of the exposition's opening in 1957 the city could claim only one car for every sixteen city residents.[16]

The expense of recasting Berlin's Hansaviertel as a *stadtlandschaft*, or "city-in-the-park," guaranteed that what was intended as a model project for inner-city reconstruction ended up a one-time performance. Modernist urban design concepts demanded a blank slate. Although extensive destruction of the site's nineteenth-century building stock provided an opportunity for experiments, free-standing housing scattered about a reconfigured traffic network meant that an intact urban infrastructure — foundations, water mains, sanitation lines, and streetbeds (all of it underground, and thus protected from the wartime bombing) — rather than conserving the construction budget through reuse, would end up generating demolition expenses. The touchy issue of expropriating private property, exacted by the wholesale reconfiguration of building plots, also pushed costs upward. West German politicians saw red when they looked at condemnation and compulsory purchase, the time-honored powers of coercion wielded by the public domain. "Berlin is compelled to promote private development," a member of Interbau's supervisory board explained, "since this must stand as a counterexample to the Stalinallee, one that documents the creative energy of private initiative as opposed to the legal caprice and collectivism of the East."[17] A way out of the dilemma was sought through the formation of a private stockholding corporation to purchase individual lots and sell the reconfigured, redeveloped property to new private and public owners, and because outright expropriation remained the tactic of last resort, and since construction costs were covered by the city and state, any victory over collectivism was entirely symbolic. Interbau apartments cost almost 60 percent more to build than typical social housing, even though rents in both were set at comparable levels through massive public subsidy. Like the Stalinallee, the reconstruction of the Hansaviertel has been regarded as a case study in Cold War profligacy, with ideological competition pushing construction costs well beyond the budget of conventional urban housing.[18] Alternatively, both can be seen as projects produced outside the economic constraints of public housing, not in spite of them. In addition to their residential functions, these fragmentary utopias were spaces for the ritual staging of new identities. In these two cities of the future, domestic practices invoked the state's prescriptions for modernity and its ideal subjects. In addition to being economically improvident housing programs, both the Hansaviertel and Stalinallee also should be considered public monuments, with correspondingly monumental budgets.

POSTWAR IDENTITIES AND THEIR TRANSMISSION BELTS

International modernism and Socialist Realism were more than just representations of the cultural hybridization fostered by the United States and the Soviet Union. The very methods by which design practices were transplanted from the superpowers to their respective fragments of Germany established parameters for new subjectivities. Modernism and Socialist Realism choreographed processes which reshaped German identity. Claims of cultural authority explicit in both aesthetics enhanced what Michel Foucault has called "governmentality" — that is, the disciplinary hierarchies of power and professed truth that condition citizens' everyday life and perceptions. East and West German identity was influenced as profoundly by the mechanics of cultural cross-pollination as by the hybridized urban environments it produced.

Socialist Realism initially was transmitted to East Germany through the dissemination of exemplars to be adopted voluntarily. Eastern German newspapers promoted the notion that Soviet art was the world's most progressive as early as 1946.[19] Few specifics were provided for the guidance of architects or urbanists, however. In spite of the implications of Soviet tirades against avant-garde art and literature, modernists in eastern Germany surveyed the prospects for reconstruction with high expectations. In Soviet-occupied Dessau, home to the battle-scarred Bauhaus building, a group of alumni regrouped with the aim of restructuring their alma mater as a laboratory for applied environmental design. Another group of modernist designers, many of them also Bauhaus-affiliated, was assembled by Hans Scharoun as members of a state think-tank on reconstruction based at East Berlin's Academy of Sciences.

Attempts to establish modernism as the architectural lingua franca of a socialist Germany were short lived, however. In April 1950, a group of six delegates from East Berlin's Ministry of Construction departed by train for Russia. Three had been trained as modernists in the 1920s; of these, two were Bauhaus alumni. In Moscow they were reschooled in the practices of Socialist Realism, the only artistic method approved by the "architect of world socialism" (and its ultimate authority on all topics), Stalin. Insurance that Soviet paradigms would be faithfully reproduced in East Germany came in the form of a typed manuscript presented to the delegates. What they had learned in consultation with Soviet advisors was condensed into three pages of Russian text organized under sixteen headings. Translated into German overnight and taken back to Berlin, the "Sixteen Principles of City Planning" passed into East German law on September 15, 1950, as the national edict on urban reconstruction.[20]

In a strategy synonymous with Stalinism, these new regulations were institutionalized in a "revolution from above." When architects failed to greet the new mandate with enthusiasm, structures of reward and punishment appeared as added incentives. East Germany's Deutsche Bauakademie, founded on January 1, 1951, copied its hierarchies of authority and party guidance from the Soviet Academy of Architects. The Bauakademie's officers were an avant-garde — in the military sense, rather than the aesthetic — in an offensive to entrench Socialist Realism in

East Germany. Kurt Liebknecht, a member of the mission to Moscow, received an instant promotion to the presidency of the Bauakademie. He introduced one of its first publications, a translation of the entry on architecture from the *Great Soviet Encyclopedia*, with the war cry: "We stand in the midst of a battle for a new German architecture."[21] This combat metaphor was more than just hollow rhetoric. Over the next few years East German architectural authorities would weave an alarming story of conspiracy and sabotage around the revival of modernism in Marshall Plan Europe. Edmund Collein, a Bauhaus alumnus who assumed the Bauakademie's vice presidency after his reorientation in Moscow, depicted the resurgence of Weimar modernism just across the border as an American plot masterminded by "monopoly capitalists" to "disassociate the people from their native land, from their language and their culture, so that they adopt the 'American lifestyle' and join in the slavery of the American imperialists."[22] Wartime bombardment had been the first phase in this blueprint for cultural imperialism, according to Collein. The follow-up was a barrage of faceless, placeless glass boxes calculated to disfigure West German townscapes and finish off whatever scraps of traditional urbanism had survived. Even more shocking was the allegation that "supporters of Bauhaus formalism," the agents responsible for despoiling West Germany's architectural heritage, were operating within the ranks of East Germany's design establishment.

The ensuing campaign to root out these enemies of the state had two outcomes. One was to establish Soviet design practice as "the great example" to be emulated in East Germany; the other was to recast social structure and identity in a Stalinist mold through responsibilities defined by class. The Pauline conversion of Hermann Henselmann from modernist to pioneering German Socialist Realist provides a case study. Henselmann was less than enthralled by the arrival of the Soviet-formulated "Sixteen Principles." His disapproval was provoked by everything from the rushed translation ("not colloquial") to the document's political overtones ("too dictatorial," and reminiscent of a style of urbanism "in a very embarrassing sense popularized by Hitler").[23] But as the cultural revolution took hold, new relationships of authority converged upon the architect from above and below to bring his behavior into line with that of a proper member of the intelligentsia. As the model subjectivities of socialism, East German workers learned that their proletarian responsibilities included assimilation of Soviet enlightenment and vigilance to assure the creation of a sound national culture. When Henselmann's modernist plans for collective farm communities were shown to East German workers late in 1950, they sent word back to the Ministry of Construction that the designs were "something insane" and that "sabotage was probable."[24] Henselmann continued to submit modernist proposals in state-sponsored design competitions, including that of the Stalinallee. With this intransigence in mind, and evidence of Henselmann's class-alien behavior established by grass-roots informants, a trap for the designer was set from above. An interviewer from the party newspaper *Neues Deutschland*, aware that Henselmann had been openly dismissive of the Socialist Realist architecture of Berlin's new Soviet embassy, asked the architect point blank for his opinion of the building.

Henselmann demurred. His reward for this attempt at diplomacy was a denunciation, entitled "On Building Style, Political Style, and Comrade Henselmann," penned by the newspaper's editor.[25] The architect's penitence came in architectural and textual renditions. Henselmann's design for Berlin's Hochhaus Weberwiese, executed days after his assault by the state press, abandoned approaches to neighborhood planning and building style demonstrated in his modernist work, and established him as the designer of East Germany's first native work of Socialist Realist architecture. His article "The Reactionary Character of Constructivism," published in *Neues Deutschland* a few months later, denounced modernism in the standardized terms of Soviet discourse, including the titular use of the Russian avant-garde as a cipher for all modernist movements.[26] The process of transmitting Soviet design standards to East Germany, in this career as well as in others, not only transformed the architecture of buildings, but also that of class as well. Socialist Realism redesigned relations of power and activated chains of accountability crucial to Stalinist society's internal guidance systems.

Similarly, U.S. influence on building culture in postwar Germany attempted to foster or repress a variety of subjectivities, but it's scattershot approach to this project reflected a muddle of conflicting and contradictory policy positions. Military occupation governance initially was determined by JCS 1067, a directive drafted in 1944. It banned direct American involvement in reconstruction as the terms of a harsh peace with Germany. Housing and other aspects of economic recovery were to be the responsibility of "the German people and the German authorities."[27] But JCS 1067 and its vindictive provisions had numerous detractors in the State and War Departments, none of them more influential than General Lucius Clay, the U.S. military governor of Germany. Clay devised policies that circumvented aspects of JCS 1067 by locating its loopholes and methodically enlarging them. His first major effort to direct the course of German urban reconstruction came in 1947, when the émigré architect Walter Gropius was contracted by the U.S. Office of the High Commissioner in Germany (HICOG) to tour the U.S. zone of occupation as a public lecturer and reconstruction consultant.

As founder of the Bauhaus, Gropius was the living legacy of Weimar modernism. He had fled Nazi Germany in the 1930s to find a welcoming new home at Harvard. There, late in 1942, with Hitler commanding most of the continent's resources and the U.S. yet to conduct a single air raid on Germany, events were set in motion which would bring the designer back to his homeland after the war. In preparation for allied victory and German de-Nazification, the Harvard Committee on Military Government was assembled under Carl J. Friedrich, a professor of government, and James B. Conant, university president (and later, during the construction of Interbau, U.S. ambassador to West Germany). The committee tapped Gropius as an authority on "reliable personalities" within Germany's architectural community, and passed his intelligence information on to government officials. Five years later, Friedrich became General Clay's special advisor on governmental affairs, and in this capacity exerted his influence to secure a position for Gropius as Clay's consultant on urbanism.[28]

Modernism's New-World naturalization was manifested in Gropius' prescriptions for German social reform through the adoption of new "democratic" homes and settlement patterns. According to the Harvard design professor, German industrial cities needed to be reformatted into "home-town" communities of 5,000 to 8,000 people, a formula conforming to the "neighborhood units" theorized by the American sociologist Clarence A. Perry. Town halls, modeled on the New England meeting house, would provide an arena for the rituals of local democracy. Housing would roll off assembly lines, just like cars.[29] Gropius' lecture tour and its public relations campaign, orchestrated by Clay's information-control officers, revived the Bauhaus in the German popular imagination as a living tradition rather than a defunct arts academy. Press releases emphasized that Bauhaus design had flourished in America, where it was considered architecture's cutting edge. In his summary report to Clay, Gropius recommended that efforts to reorganize "the 'Bauhaus' movement" be supported by the military government.[30] However, not even friends in high places could bring Gropius' vision of postwar German reconstruction to fruition. Its synthesis of modernist architecture, suburban American settlement patterns, and high-tech construction technology might have been suited to a new-towns program in an industrially developed country, but it was a poor match for Germany, which was struggling to reconstruct its cities with an economy equally in ruin.

The drive to transplant values and identities associated with American democracy into post-Nazi society also pervaded U.S. efforts to re-educate German urban planners. A 1949 pilot program sponsored by HICOG brought three highly placed German city planners to America for retraining. This trial run was judged a failure. Upon return to their homeland, German converts to "the American way of planning" found themselves shut out of positions of power by conservative colleagues. Hans Blumenfeld, a program evaluator hired by the State Department, concluded, "We may have been rather naïve in assuming that a visit to the U.S. would automatically increase the prestige of the visitor."[31] A second try in the spring of 1950 — just as East Germans were learning the lessons of Socialist Realism in Moscow — marked a change of strategy. Rather than a "revolution from above," the democratization of West German urban planning was to come from below, generationally speaking. Eleven participants fresh out of German architecture schools underwent a semester of intensive retraining at the University of North Carolina. Their reorientation program regarded city planning as a vehicle for grass-roots political participation and the development of a socially engaged citizenry. Here again U.S. paternalism stumbled on its own resistance to authoritarian methods and German bureaucrats' abhorrence of innovations sponsored by foreign powers and designed to transfer municipal power to the public at large. The second exchange program proved no more effective than its predecessor in placing trainees in positions of influence, amending West German urban planning practices, or creating the democratic activism that was the project's underlying rationale.

U.S. policy on German reconstruction came full circle with the European Recovery Plan (ERP). Unveiled in June 1947 at a Harvard commencement address, the ERP, better known as the Marshall Plan, was the blueprint for a unified Europe composed of liberal democracies linked by open markets — "a lubrication system for global capitalism," as one Cold War historian described it.[32] Membership in the ERP was voluntary. Member states were expected to achieve political and economic stability as interdependent trading partners. In the quest for higher productivity, Marshall Plan programs proposed injecting American building practices into West Germany, marking a complete break with the prohibitions encoded into JCS 1067. The tradition-bound German construction industry required an overhaul, according to U.S. advisors attempting to resolve the nation's mounting housing crisis. They were frustrated by resistance to the notion of planned obsolescence and the cost-saving, light-weight materials it made use of. "From the American point of view, the modest German dwelling is an expensive, solid looking, inconvenient affair. It will last a couple of centuries, and there are houses everywhere of identical construction and of just such age to prove it," wrote Vernon DeMars, a Californian managing Marshall Plan residential construction.[33] A program announced in 1951 supported experiments in housing design, and attempted to depose the obsolescent bureaucracy that stood in their way. A previous scheme aimed at inducing dogmatically conservative building officials to abandon traditional (and by American standards, hopelessly inefficient) construction requirements involved an educational film featuring the "reactionary builder type" as a caricature to be ridiculed.[34] In the Marshall Plan experimental housing program, the motivating power of cash superseded indirect prodding. Residential construction funds totaling 37.5 million DM were divided among municipalities which agreed to permit deviations from local building codes and zoning ordinances. Fifteen model neighborhoods were eventually built for about 15 percent less than housing raised by traditional construction techniques.[35] But this success proved difficult to replicate without monetary incentives to purchase opportunities for innovation.

Ruhr Valley coal miners were the beneficiaries of a second generation of German housing built under the supervision of American architects, this time as consultants to the Marshall Plan's Mutual Security Agency (MSA). Novel housing prototypes and materials were again elements in the experiment; but here the major innovation was financial. MSA advisors believed that the consumer-driven economy proposed by the Marshall Plan had to begin at home. The Ruhr program introduced mortgages which allowed working-class families to purchase a home for about the same monthly cost as a typical rent. "This must be seen in the light of developments in the United States under the FHA plan, in contrast with the previously accepted dogma in Germany that the workman must not be saddled with home ownership," American consultants maintained.[36] Received wisdom about German domestic tradition was also jettisoned. Advisors pushed for built-in kitchens and electric appliances over objections that "the most prized possession of the miner's family was the great polished stove . . . which the family carried with them wherever they moved."[37] An emblematic tool of U.S. postwar culture, the opinion survey, came to the rescue. It revealed that miners were "open to new ideas," and "desirous of improved living conditions and a more modern way of life." Although modern kitchens were expensive relative to a miner's income, American advisors asserted that "stimulating a demand for certain household equipment would be the first step to encourage more production and thereby lower unit costs for these items."[38] A traveling exhibit used pictograms to pitch the advantages and responsibilities of home ownership to miners, and displayed scale models of the proposed houses. These technologies of persuasion, combined with advantageous financing terms (reflecting reduced building costs achieved through efficient planning and competitive bidding), overturned convention. "Suspicious at first, the miner and his wife gradually warmed up to the idea, listened attentively to the architect's lecture an what this program was all about," the Americans observed, "and in some cases came back two and three times to hear exactly how much the 10% down payment and the monthly payments for which he could buy his house would be." After the exhibit visited eight communities targeted for housing construction, applicants exceeded the number of available houses by many times.[39]

The economic and ideological implications of this program went far beyond improving workers' living conditions. The miners' adoption of consumption patterns typically associated with middle-class domesticity presaged the triumph of Marshall Plan strategies for Western European reconstruction. Stalin once joked that, aside from the A-bomb, "the main armament of the Americans" consisted of cigarettes, nylons, and other "goods for sale."[40] American advisors in the Ruhr demonstrated that a different commodity — housing — was much more effective at eroding the underpinnings of a proletarian identity: namely, alienation from the means of production and its rewards. From a Marxist point of view, preconditions for a socialist revolution were sabotaged by the Ruhr project as it transformed capitalism's discontents into its consumers and homeowners — subjects with a stake in the system. Propaganda distributed by the U.S. Agency for International Development, in its invocation of residential instability as a catalyst for a communist subjectivity, suggests that housing was part of a calculated tactic for ideological containment. In 1952, a press release on Germans who had yet to experience the nation's *Wirtschaftswunder* featured a photograph titled "A Potential Danger" **(Figure 8.4)**. It showed a man with a bowl of food and a woman looking on from the window of a wooden trailer. Its caption read:

> Proudly standing behind his rickety old car in front of the caravan which is his home, this German is one of the millions who today — seven years after the end of the war — still live in discomfort and insecurity. . . . But this man, taken as a type, is more than just a name on a waiting list — he represents a potential danger to Germany, for in the years of this type of existence he has developed a taste for insecure living and has acquired a mentality which is an easy prey for communist propaganda.[41]

Cultivation of a taste for the stability of bourgeois domesticity was one remedy for the potential dangers of a proletarian identity. American-sponsored exhibitions supported this undertaking, and in the process, promoted modernism's International Style as the aesthetic analogue of Marshall Plan modernization.

Figure 8.4. "A Potential Danger . . ." This 1952 photo, taken by the U.S. Agency for International Development, was one of the many press releases designed to promote the Marshall Plan's European economic recovery programs. (Courtesy of National Archives at College Park, RG 286MP Ger 1695.)

While on their Western European circuit, exhibits of household progress *à la américaine* invariably stopped over in West Berlin, where consumer seduction was used to encourage the development of middle-class identities in the city's western sector, while subverting the socialist subjectivities under construction in the East.

GLOBAL HYBRIDITY AT MULTIPLE SCALES

The design of cities and of household furnishings were closely related in the discourses of modernism and Socialist Realism. Success for the two movements was measured in terms of universal dissemination in two senses: both across national boundaries and throughout the entire realm of cultural production. This was a defining characteristic of the "International Style," as popularized by New York's Museum of Modern Art (MoMA) in a 1932 exhibit. Curators Philip Johnson and Henry-Russell Hitchcock attempted to relieve European modernism of any trace of leftist radicalism in their pursuit of a seemingly apolitical avant-garde aesthetic. "This contemporary style, which exists throughout the world, is unified and inclusive, not fragmentary and contradictory like so much of the production of the first generation of Modern architects," they declared in the exhibit's companion volume.[42] Socialist Realists came to parallel conclusions about the totalizing nature of aesthetic reform. On the eve of postwar reconstruction, the Soviet Minister of Architecture, Arkadii Mordvinov, proclaimed aesthetic integration of every aspect of socialist life as the ultimate goal of Soviet design. "Clothes, furniture, equipment, interior decoration, buildings, streets, parks and towns as a whole — everything should be beautiful."[43] Given the Socialist Realist connotations of the word "beauty" — a term loaded with the portent that "function" held within modernist discourse — this pronouncement was nothing less than a Stalin-era analog of modernism's ideal of the *gesamtkunstwerk*, the total work of art.[44] The emphasis, shared by modernism and Socialist Realism, on a unitary aesthetic for the entire human landscape, from the city to individual housing units and their appointments, gave both movements a quality which scientists call "similarity across scale" — that is, patterns that persist when viewed from a distance, close-up, and anywhere in between. Along with urban prototypes like the Stalinallee and Hansaviertel, modernist and Socialist Realist paradigms for domestic interiors provide a parallel perspective on the Cold War's processes of hybridity and cultural globalization.

America's first postwar intervention in Western European household design was the 1951 "American Home Furnishings" exhibit. It brought a U.S. variant of Bauhaus design to Germany (among other countries), courtesy of the U.S. State Department and the ERP. These government agencies commissioned MoMA's curators in New York to assemble a collection of domestic articles which embodied "American design and craftsmanship as adapted to American home living."[45] However, this display of transatlantic domesticity was anything but representative. The period and regional nostalgia that dominated U.S. taste was nowhere to be found in the show. Instead, 500 objects exemplary of "progressive American design" and characterized by "simple lines and functional forms" taught how a combination of "many low-cost items and a few luxury items complemented a contemporary lifestyle."[46] In Berlin, this display of domestic modernity was staged in the new Marshall House, a pavilion built by the U.S. in 1950, and located within easy reach of day-tripping East Berliners. At a time when raw utility dominated new construction, the Marshall House's revival of a luxurious variant of Weimar-era modernism caused a sensation among local architects. For West and East Berliners alike, the equation of Bauhaus-inspired design with American cultural authority was hard to miss.

The "American Home Furnishings" exhibition raised sticky questions about Marshall Plan design policy. Its calculated omission of the diversity of objects produced for a consumer marketplace characterized by "freedom of choice" simulated the hegemonic aesthetic regulation advocated by Socialist Realism. What could account for the exhibition's misrepresentation of the U.S. domestic design as an artifact of something other than free-market capitalism? Although American authorities may have wanted to impress foreign elites with the sophistication of U.S. culture by adopting "highbrow" standards, workers were the main target audience for Marshall Plan propaganda, including its exhibitions. And in promoting the delights of consumer capitalism among the working class, goods that were pretty, rather than austere, might well have won over converts more effectively.

The broader ideological and economic strategies behind the U.S. sponsorship of modernism become clear only when the "American Home Furnishings" show is considered in concert with a subsequent exhibition. "We Build a Better Life" opened at Berlin's Marshall House late in 1952 and toured Western Europe the following year. Sponsored by MSA, HICOG, and the West German government, this exhibit's theme was the higher standard of living that could be attained through the Marshall Plan goals of increased productivity, abolition of trade barriers, and economic integration of "the peoples of the Atlantic community." The main attraction at this commodity spectacle was an "ideal house," a single-family dwelling realized down its to kitchen gadgets and garden tools, but built without a roof so that visitors could get a bird's eye view of interiors from an overhead catwalk. This stage-set for the domestic life of "an average skilled worker and his family" was manned by models enacting household tasks and leisure rituals in a consumer wonderland. The exhibit was a hit among visitors, half of them from East Berlin, but it sparked an internal controversy among American officials. MSA housing advisors objected to an "ideal house" with twice the square footage of the legal average, as stipulated in West Germany's egalitarian housing law. Marshall Plan administrators responded that "the exhibition was not about housing at all. Its main point was the attractive and realistic display of the least expensive, aesthetically acceptable mass-produced objects commonly used in everyday living by ordinary people."[47] These furnishings could be examined at close range in another part of the exhibit, introduced by a billboard featuring the photograph of a male worker with the headline: "This man is a worker and at the same time a consumer" **(Figure 8.5).** Every object displayed was contemporary in design, manufactured in a Marshall Plan nation, and bore a tag indicating origin, retail price, and the number of hours of labor by a skilled worker needed to purchase it. As a primer to "the modern approach to interior decoration," the lesson was that "rationally designed products from different countries in the Atlantic community can be combined harmoniously."[48] International modernism, as envisioned in the "Better Life" exhibit, represented both a mode of global capitalism and its aesthetic. As retooled by the Marshall Plan, MoMA's formula for modernism provided aesthetic redemption for bourgeois domesticity, purging it of nineteenth-century fustiness and obsolete class associations, and

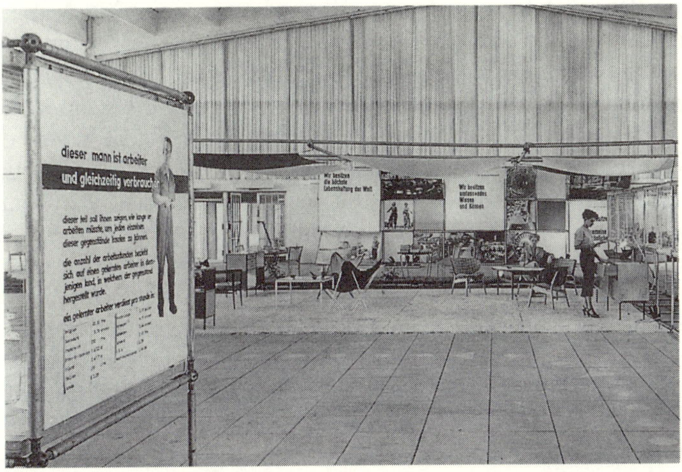

Figure 8.5. An international display of domestic furnishings at the "We Build a Better Life" exhibition. Visitors are welcomed by the billboard on the left with the headline, "This man is a worker and at the same time a consumer." After opening in conjunction with an industrial fair in West Berlin in 1952, the show went on to Stuttgart, Hannover, Hamburg (where this photo was taken), and other Western European cities. (Courtesy of National Archives at College Park, RG 286 MP Gen 1974.)

recalibrated its household artifacts to a global paradigm of mass consumption premised on the erosion of trade and cultural barriers.

Varied meanings accompanied this cultural strategy at its various scales of reference. Within West Germany, the language of modernism embodied in MoMA's "International Style" had a specific utility. Its local dialect was Bauhaus modernism, revitalized by American patronage. The revival of Weimar modernism symbolized continuities between the liberal democracy of the interwar Weimar Republic and the postwar Federal Republic, and created a shared turf where American and West German paradigms of cultural modernity could meet and mingle.[49] Across Western Europe, Marshall Plan exhibitions presented International Style modernism as an aesthetic suited to the open markets seen by American policy-makers as the surest path to postwar harmony. There is no better indicator of West Germany's embrace of this transnational model of material culture than its rendition at metropolitan scale in West Berlin's Interbau: a collection of buildings designed and outfitted by architects from a dozen "free-world" nations and offered as a prototype for postwar urban reconstruction featuring modernism's "*genius loci* from everywhere."[50]

The reinvention of international modernism as the common culture of a capitalist trade block was reinforced by Socialist Realist reaction, as demonstrated by an East German counteroffensive on the interior design front launched from East Berlin's "House of Culture of the Soviet Union" in March 1952, a year after the opening of the "American Home Furnishings" show. Under the auspices of the Bauakademie, hundreds of architects and designers met with furniture manufacturers and representatives of state retailing organizations to discuss socialism's domestic appointments. Minister President Walter Ulbricht proclaimed in a keynote address that Berlin was "more beautiful than ever before, as demanded by the will of the people." Bauhaus-style furniture was incompatible with this vision, because it failed to live up to "the aesthetic experiences of the new Germany's progressive human beings." East German designers and manufacturers too often rationalized the production of "crate-like" furniture of "primitive design" as an economizing measure, Ulbricht lamented. Even the conception of the kitchen as a "workshop" was "formalistic," according to Ulbricht. Housewives wanted beauty, not cabinet doors that were as flat and featureless as walls. Workers demanded furniture that was cheap, of course, but comfortable, useful and beautiful as well. It was high time for designers to establish closer ties to their working-class clients through systematic study of local tradition, according to conference speakers. The "great tasks of interior design in the battle for a new German culture" would only be resolved by rediscovering progressive qualities embodied in the national craft heritage, and repudiating what a Bauakademie spokesman called the "colorless international language" at the root of the "so-called American lifestyle."[51]

Research into precedents for Socialist Realist interior design intensified in the summer of 1952, just as Ulbricht initiated an all-out effort to "construct socialism." East German photographers traveled to dozens of regional museums to document collections of furniture, lighting fixtures, and textiles. They concentrated on Neoclassical objects dating from the early nineteenth century, "the last great epoch of unified German decorative and domestic culture."[52] Their efforts yielded an authoritative catalog of source material for the "appropriation and creative development" of German tradition in the decorative arts.[53] But this publication, four years in the making, was obsolete the moment it appeared. Ulbricht's crash program to reproduce the Stalinist system in Germany had left society demoralized and the economy in tatters. As central planners stoked heavy industry, consumer goods all but disappeared, and those which could be found were usually shoddy and expensive. A disenfranchised proletariat revolted in June 1953, staging street protests in cities throughout East Germany. In view of its catastrophic context, the quest to reinvent German tradition in the service of socialist domesticity was a chimera. Presented with the choice of Socialist Realism's brand of anti-imperialism or the "slavery of the American lifestyle" ascribed to the Marshall Plan, citizens of the East voted with their feet. In the first four months of 1953 alone, some 120,000 East Germans fled into exile, most across the urban border into West Berlin.

Stalin died in March 1953. A year later his successor, Nikita Khrushchev, galvanized Moscow's architectural establishment with a fierce denunciation of the Soviet construction industry's shortcomings. Khrushchev's initiative "to end the country's housing shortage in 10-12 years" dismantled the ideological underpinnings of Socialist Realist representation in architectural design.[54] A 1955 central committee resolution "on the removal of exaggerations in planning and building" proposed the development of a new building culture devoid of "contrived, expensive decoration," and instead characterized by "simplicity, austerity of form, and economy of layout."[55] Architects were instructed to abandon the autarkic worldview cultivated under Stalinism, and look to foreign achievements for inspiration, including those of the capitalist West. A new venture in globalization and urban hybridity was about to begin.

A POST-STALINIST POSTSCRIPT

Among Interbau's contributions from across the "free world," Le Corbusier's design for an *unité d'habitatation* provoked controversy the moment plans were unveiled. The building was too big to fit on the Hansaviertel site. Ceiling heights in this *machine à habiter*, it was said, were too low to work in Berlin, where men and women were on average ten centimeters taller than in Marseilles. The architect dismissed such objections with a breezy affirmation of the universality of his brand of modernism. "My buildings are good for Zulus and Eskimos," Le Corbusier declared. "They are neither dependent on the height of people nor on century-old cultural structures."[56] But were they, along with the rest of Interbau's housing experiments, compatible with the new structures and people of socialism? Three members of the Bauakademie set out to answer that question by quietly crossing the East German border to merge with the crowds at the 1957 exposition.

Direct observation and a survey of Western architectural literature yielded an analysis of Interbau's relevance to the Bauakademie's new assignment — industrialized housing. According to its socialist assessors, Interbau not only exhibited new trends in construction, but also all of the shortcomings of capitalism as well. Aesthetically, the Hansaviertel was a monument to uncontrolled individualism. The "outspoken emphasis on individualistic peculiarities" in work by prominent architects, and a muddled mix of short and tall buildings, revealed an aversion to "all regularity and visible order." Still, the attempt to satisfy an "indistinctly defined 'modern lifestyle' and its domestic requirements" had produced some positive results. Further development was deemed necessary, because "the propertied class" and "the 'well-situated' petty bourgeois" were the intended beneficiaries of Interbau's "endeavors to isolate the family within its own home." The East German research team rejected as sheer propaganda claims that Interbau provided a test bed for new concepts in social housing. "Efforts made in recent years by the Federal Republic to support private housing under any circumstance and at the expense of co-operative construction has led to a relative decline in the proportion of social housing." In the realm of construction technology, Bauakademie observers concluded that West Berlin had spent enormous sums to corroborate what Soviet research had already demonstrated: that replacing manual workmanship with industrially prefabricated building components could yield dramatic reductions in the cost of new construction.[57]

The East Bloc's repudiation of Socialist Realist urbanism signaled a new imperative to compete with capitalism's global economy and culture on some of its own terms. Housing projects assembled from precast concrete parts soon finished off the last stretch of the Stalinallee, which exchanged its embarrassing name for the more seemly "Karl-Marx-Allee." In scrapping the ideology and iconography that had distinguished socialist urbanism from its capitalist competition, Khrushchev placed their products on a convergent course, inviting direct comparisons. Districts of paneled slabs rising in the center of East Berlin were planned in accordance with "the best neighborhood principles," according to a British journal, but they seemed to "completely reject the ideas of city scale and city grandeur, which were the saving grace of the symmetrical Stalinallee."[58] Large-scale, rationalized construction technology was indeed more efficient than handwork, but rather than evading the issues of architectural representation that were at the heart of Socialist Realism, this alternative posed dilemmas which were to remain unresolved. Stalinism's grotesque prescriptions for a civic utopia, once abandoned, opened the way for whole residential suburbs to go up quickly, and at great savings. This enrichment of the housing stock in East Bloc cities, however, came at the expense of having them increasingly look like the poor relations of their equivalents in the West, rather than the artifacts of an altogether different civilization.

NOTES

1. For a brief history of the evolution of Socialist Realism in architecture as a medium of national identity, see G. Castillo, "Peoples at an Exhibition: Soviet Architecture and the National Question," in T. Lahusen and E. Dobrenko (eds.), *Socialist Realism without Shores*, Durham, NC, University of North Carolina Press, 1997, pp.91–119.

2. For a discussion of the Cold War superpowers as "informal empires," see J.L. Gaddis, *We Now Know: Rethinking Cold War History*, New York, Oxford, 1997, pp.26–53.

3. B. Miller Lane, "The Berlin Congress Hall, 1955–1957," *Perspectives in American History 1* (new series), 1984, p.139, fn.22.

4. B. Flierl, "Stalinallee in Berlin," *Zodiac* 5, 1988, p.79.

5. H. Riecke, *Mietskasernen im Kapitalismus, Wohnpaläste im Sozialismus*, Berlin, Kultur und Fortschritt, 1954, p.7.

6. The construction budget for Stalinallee apartments was set in 1952 at an average of 40,000 marks per housing unit (in 1950 marks). In 1950 the state's target cost for new urban housing was 10,000 marks per unit. H. Nicholas and A. Obeth, *Die Stalinallee: Geschichte einer deutschen Straße*, Berlin, Verlag für Bauwesen, 1997, p.145.

7. K. Mahler, "Internationale Bauaustellung 1956: Wiederaufbau eines inneren Stadtviertels," *Bauwelt* 35, 1953, p.681.

8. G. Suter, "Ein Exposé für Berlin: Gespräche mit einer Idee," *Die Neue Zeitung: die amerikanische Zeitung für die deutsche Bevölkerung*, November 15, 1953; reprinted in J.F.

Geist and K. Kürvers, Das Berliner Mietshaus, 1945-1989, Munich, Prestel, 1989, pp.366–367.

9. V.M. Lampugniani, "The 'Zero Hour': Reconstruction's Goals and Premises," *Domus*, vol.21, 1987, p.96.

10. T. Heuss, in Internationale Bauaustellung Berlin (ed.), *Interbau Berlin* 1957, Berlin, p.13.

11. D. Howell-Ardila, "Berlin's Search for a 'Democratic' Architecture," *German Politics and Society 48*, vol.16, no.3, 1998, p.72.

12. K. Liebknecht, *Die nationalen Aufgaben der deutschen Architektur*, East Berlin, Deutsche Bauakademie, 1954, p.74.

13. W. Gropius to O. Bartning, December 28, 1955, Reginald Isaacs Papers, Smithsonian Archive of American Art, Washington, DC.

14. R. Rosner, "Exhibition, or New City?" *The Architect's Journal*, August 15, 1957, vol.126, p.242.

15. H. Bodenschatz, "Antworten West-Berlins auf die Stalinallee," in H. Engel and W. Ribbe (eds.), *Karl-Marx-Allee Magistrale in Berlin: Die Wandlung der sozialistischen Prachtstraße zur Hauptstraße des Berliner Ostens*, Berlin, Akademie Verlag, 1996, pp.161–162.

16. H. Stephan, "Der Hansaviertel in der Gesamtplanung von Berlin," in Internationale Bauaustellung Berlin (ed.), *Interbau Berlin* 1957, p.22.

17. H. Bodenschatz, "Antworten West-Berlins auf die Stalinallee," in Engel and Ribbe (eds.), *Karl-Marx-Allee*, p.155.

18. Ibid., pp.159–160.

19. D. Pike, *The Politics of Culture in Soviet-Occupied Germany, 1945–1949*, Palo Alto, CA, Stanford University Press, 1992, pp.226–230.

20. For the history and archival documentation of this retraining mission, see Institut für Regionalentwicklung und Strukturplanung (ed.), *Reise nach Moskau: Quellenedition zur neueren Planungsgeschichte*, Erkner, IRS, 1995.

21. B.P. Michailow, *Grosse Sowjet — Enzyklopädie: Architektur*, Berlin, Bruno Henschel und Sohn, 1951, p.5.

22. E. Collein, "Die Amerikanizierung des Stadtbildes von Frankfurt am Main," *Deutsche Architektur*, vol.1, no.4, 1952, pp.151–155.

23. Institut für Regionalentwicklung und Strukturplanung (ed.), *Reise nach Moskau*, pp.179–181.

24. Ministerium für Aufbau, Arbeitsprechung, October 10, 1950, DH1.44468, Stiftung Archiv der Parteien und Massenorganisationen der DDR im Bundesarchiv (Aussenstelle Coswig).

25. R. Herrnstadt, "Über den Baustil, den politischen Stil und den Genossen Henselmann," *Neues Deutschland*, July 29, 1951. Reproduced in Andreas Schätzke, *Zwischen Bauhaus und Stalinallee: Architekturdis-kussion im östlichen Deutschland 1945–1955*, Braunschweig, Vieweg, 1991, pp.140–143.

26. H. Henselmann, "Der reaktionäre Charakter des Konstruktivismus," *Neues Deutschland*, December 4, 1951.

27. C. Eisenberg, *Drawing the Line: The American Decision to Divide Germany, 1944–1949*, Cambridge, Cambridge University Press, 1996, p.47.

28. C.J. Friedrich to C. D'Arms, July 22, 1947, Clay file, Reginald R. Isaacs papers, Smithsonian Institution Archives of American Art, Washington, DC.

29. "Gropius Maps a Germany of New England-Type Communities," *Boston Globe*, January 1, 1948.

30. W. Gropius, "Report of Dr. Gropius," September 5, 1947, 250/83/24/3, Box 49, Lot 5GF99, HICOG, Office of the Executive Secretary, General Records 1947–1952, Record Group 466, National Archives at College Park, College Park, MD.

31. H. Blumenfeld, "City Planning in Germany," undated mimeograph (author's collection).

32. Gaddis, *We Now Know*, p.194.

33. V. DeMars and W.K. Wittausch, "German Coal Miner's Housing Program Progress Report," 1952, p.11, 250/79/28/03–4 box 3, Records of the U.S. Foreign Assistance Agencies, Housing 1951-2, Special Representative in Europe, Office of Economic Affairs, Labor Division, Record Group 469, National Archives at College Park, College Park, MD.

34. J.M. Diefendorf, "America and the Rebuilding of Urban Germany," in Diefendorf, A. Frohn, and H.J. Rupieper (eds.), *American Policy and the Reconstruction of Germany, 1945–1955*, New York, Cambridge University Press, 1993, p.325, n.138.

35. J.M. Diefendorf, *In the Wake of War: The Reconstruction of German Cities After World War II*, New York, Oxford University Press, 1993, pp.140–143; B. Wagner, "German Architecture Looks Up: ECA Housing Development Helps German Architects and Contractors to Introduce Modern Methods in Housing," *Journal of the American Institute of Architects*, 1952, vol.17, pp.13–14; and H. Wandersleb and H. Schoszberger (eds.), *Low Cost Housing in Germany*, Bonn, Ministry for Housing of the Federal Republic of Germany, 1952, pp.1–4.

36. DeMars and Wittausch, "German Coal Miner's Housing," pp.11–13.

37. Ibid.

38. Ibid.

39. Ibid.

40. The comment was made by Stalin to Zhou Enlai in 1952. V. Zubok and C. Pleshakov, *Inside the Kremlin's Cold War: From Stalin to Khrushchev*, Cambridge, MA, Harvard University Press, p.75.

41. Photograph Ger 1695, "A Potential Danger . . . ," 1952, Records of the Agency for International Development, Record Group RG 286, National Archives at College Park, College Park, MD.

42. H.R. Hitchcock and P. Johnson, *The International Style*, New York, Norton, 1995, p.35–36.

43. A. Mordvinov, comments at the Sixth Session of the Soviet Academy of Architecture, December 1943, as quoted in A. Tarkhinov and S. Kavtaradze, *Stalinist Architecture*, New York, Rizzoli, 1992, p.98.

44. The concept of beauty in Soviet discourse is discussed by L. Heller, "A World of Prettiness: Socialist Realism and its Aesthetic Categories," in Lahusen and Dobrenko (eds.), *Socialist Realism without Shores*, pp.51–75.

45. P. McCray, "American Home Furnishings Exhibit in Stuttgart" (press release), May 25, 1951, Gen 893–1012, Records of the Agency for International Development, Record Group RG 286, National Archives at College Park, College Park, MD.

46. Ibid.

47. M. Harris to J.M. Heath, December 4, 1952, 250/79/28/03–04, Box 3, Records of the U.S. Foreign Assistance agencies 1948-61, Special Representative in Europe, Office of Economic Affairs, Labor Division, Country Files Related to Housing 1951-52, Record Group 469, National Archives at College Park, College Park, MD.

48. Photographs Gen 1974 and Ger 2219–2226, "Productivity and Integration Make For Higher Standard of Living," March 1953, Records of the Agency for

International Development, Record Group RG 286, National Archives at College Park, College Park, MD.

49. P. Betts, "The Bauhaus as Cold War Legend: West German Modernism Revisited," *German Politics and Society* 2, 1966, p.76.

50. W. Sewing, "Genius loci von überall: Lokale und internationale Baukultur in Berlin," *Foyer* 2, 1996, pp.32–35.

51. "Bericht über die Arbeitstagung der Instituts für Innenarchitektur der Deutschen Bauakademie über Fragen der Innenarchitektur und des Möbelbaues am 14 March 1952," Institut für Innenarchitektur, Deutsche Bauakademie, DBA/DH2/DBA/A141, Stiftung Archiv der Parteien und Massenorganisationen der DDR, Bundesarchiv Berlin.

52. Skujin Files, Institut für Innenarchitektur, Deutsche Bauakademie, DBA/DH2/VI/07/29, Stiftung Archiv der Parteien und Massenorganisationen der DDR, Bundesarchiv Berlin.

53. G. Wernitz (ed.), *Historische Möbel und Innenräume*, Berlin, Henschelverlag, 1956.

54. M. Heller and A.M. Nekrich, *Utopia in Power*, New York, Summit, 1986, p.559.

55. A. Åman, *Architecture and Ideology in Eastern Europe during the Stalin Era: An Aspect of Cold War History*, Cambridge, MA, MIT Press, 1992, p.219.

56. W.J. Siedler, "Città e Utopia," *Domus*, no.685, 1987, p.22.

57."Analyse der 'Interbau'," Deutsche Bauakademie, DBA/DH2/A/48, Stiftung Archiv der Parteien und Massenorganisationen der DDR, Bundesarchiv Berlin.

58. S. Rosenberg, "Berlin and the 'Hauptstadt Berlin' Competition," *Architects' Yearbook 9*, 1960, p.75.

9

Porous Boundaries: Fence Patterns and Mexican-American Identity in San Antonio, Texas

Robert Mugerauer

The concept of the "hybrid" is emerging as a triply important metaphor in current attempts to describe and interpret built environments. First, because the word connotes and denotes the organic realm, it provides a useful means to generate counter-analyses to those produced by earlier mechanical ways of thinking. Secondly, the term provides a name for the offspring of unions of things that share similarities while having important differences, and it involves a process wherein first-generation similarities and differences somehow persist, albeit in modified form, in the hybrid (in short, it is precisely in the hybrid that socially important differences and multiplicities in processes of change occur). Thirdly, nature and nurture occur together in the hybrid: thus, in deliberately producing a third entity from two others with similarities and differences, societies may identify, acknowledge and respond to the characteristics that matter to them and (at least implicitly) take responsibility for the ways they modify or reconstitute the world.

The key aspects of the hybrid — an organic worldview that acknowledges and responds to both natural and socially constructed similarities and differences — obviously have much to offer researchers now that the social construction of group identity and its physical settings have reemerged as a central theme in environmental research. Contrasts between mechanically oriented modes of analysis and organic ones are striking, as can be seen by comparing typical texts that utilize, respectively, metaphors of the mechanical and the hybrid. Oscar Newman speaks clearly in the first voice:

> We are witnessing a breakdown of the social mechanisms that once kept crime in check and gave direction and support to police activity. . . . We have become strangers sharing the largest collective habitats in human history. . . . "Defensible space" is a surrogate term for the range of mechanism — real and symbolic barriers, strongly defined areas of influence, and

improved opportunities for surveillance — that combine to bring an environment under the control of its residents.[1]

Compare this with Norma Evenson's use of biological metaphors of propagation:

> Bombay, Madras and Calcutta were physical and social hybrids virtually from their inception. . . . They embodied mixtures of population groups not found elsewhere, with unique juxtapositions of caste, religion, and language. . . . The unprecedented size of the metropolis, with its rapidly-evolving fabric, promoted new building types and constant alteration of traditional building forms.[2]

As to issues of differences and similarities of groups and environments, these clearly appeared in the socially focused behavioral and perceptual studies of the 1960s and 1970s, which analyzed the physical demarcation of territories, the cultural enforcement of surveillance and safety, and the affirmation of personal identity and responsibility in settings such as public housing projects. Although this work still is appreciated and remains inspirational because of its sensitivity to cultural difference and its often fine-grained case-study method, it now also appears somewhat politically or ideologically "innocent." These earlier researchers, especially as sociologists and anthropologists, did appreciate the struggles of displaced or ethnically diverse groups; but, in the aftermath of poststructuralism and cultural studies of the last decade, most of the work of the 1960s and 1970s now seems to have too easily assumed the neutrality and givenness of the phenomena of identity and environments.

The shift apparent here is not merely a matter of the contrast between positivistic methodology and the historical, social analysis provided by forms of Marxism, but also includes a methodological change from the universalizing outlook of objective rationalism to the more localized strategies of critical theory. Thus, returning today to issues of identity and environment, researchers contend that social-political identities are neither neutral nor given — to be "objectively represented" or "expressed" — but are themselves socially produced and historically contested. Seen in this manner, the questioning of relationships between ethnic identities and physical environments involves rendering explicit the open, historical dynamics among competing groups or factions — a process involving the simultaneous and interdependent constitution of identities and places that is carried out, for example, by way of the production of private and public spaces and buildings, and through the uses of differentiated forms, materials, and modes of construction.

EARLY RESEARCH AND PRACTICE

The first wave of research and practice in the 1960s and 1970s remains known through the still-influential work of figures such as Paul Oliver, Edward T. Hall, Clare Cooper-Marcus, Jane Jacobs, Oscar Newman, Robert Sommer, and many others.[3] These pioneering figures did fieldwork in highly diverse settings around the world, and they emphasized the social factors involved in the development of community forms and built environments. As a result, their theorizing and practices emphasized the dynamic manner in which inhabitants of settlements or participants in housing design and building manifest their identities. In this first wave, Paul Oliver argued that there are many social aspects,

> . . . that seek or invest meaning in the structures of vernacular architecture. Their purposes are as varied as the multitudinous forms in which they gain expression. Together they present a spectrum of rites, customs, superstitions, religious practices, signs, two and three-dimensional symbols, hierarchies of space, patterns of behavior, expressions of built form and other aspects of symbolic usage which centre upon the building, the village or settlement. . . .
>
> We may eventually encourage the expression of group identity through symbols in vernacular architecture, rather than repressing it on grounds of taste, bye-laws or planning regulations.[4]

This research of the 1960s and 1970s, and the continuing work elaborating and refining the basic findings, established that residents of standardized dwellings, including public housing, tend to — and should be encouraged to — personalize their units (to the extent allowed) in ways that manifest their cultural heritages and the degree and manner in which they are, or desire to be, assimilated (or not) to the dominant surrounding culture.[5] The largest part of the research literature focused on major symbolic and functional elements (especially roofs, doors, porches, ornamental and decorative features, plantings and gardens, courtyards, plazas, and play areas) when describing similarities, differences and changes between the environments and identities of specific ethnic groups and those of the surrounding areas.[6] The same was true of literature that analyzed the ways public housing should or should not be related to its context.[7]

Yet even as established humanistic methods of theory and case study documented and described the cultural and material conditions and expressions of sense of place, as Dennis Crow has argued they also tended to leave out consideration of the multiple and contested dimensions of place and identity.[8] It seems an obvious enough and complementary point: place and identity are neither neutral nor universal, but are constituted through the productive practices of specific historical groups.[9] In other words, the process is fundamentally political, and as the groups work out their own differentiated as well as commonly shared spaces and identities, varying strategies and degrees of power come into play, so that some groups' identities and places dominate while some recede, and so that some environmental patterns are richly polysemous and participatory whereas others are reductive and exclusionary.

Hence, more recently, anthropological and sociological work on related themes has continued to emphasize case study and cross-cultural comparison, but has pushed understanding of the extent to which identities are fluid and contested, rather than stable and more "simply" marked or asserted.[10] As one example at the largest scale, arguments have been made that entire natural-cultural landscapes need to be seen in terms of contested power relations.[11] The empirical question of how place and identity are socially constituted, as Lefebvre has shown, cannot be reduced to a conceptual consideration of mental space, as Derrida and other poststructuralists have argued; rather, the focus should be on the built environment. According to Keith and Pile:

> Space cannot be dealt with as if it were merely a passive, abstract arena on which things happen. The spatialities of urban regeneration and the politics of diaspora illustrate precisely these themes.... The first may be understood as an identity politics of place and the second as the spatialized politics of identity.[12]

Thus, in their current attempts to find a more positive alternative to deconstruction, one that still holds honestly to the critical tenets necessary for thinking and work today, environmental researchers are shifting to politically informed analysis of the differences at play in various localized formations of identity and place. Such is the case with the present description and analysis of Mexican-American identities gained, and then lost, within San Antonio's Mirasol Homes. As feminist critic Angelika Bammer has explained:

> It is not surprising that displacement has played such a prominent role in the operative theoretical paradigms with which we have attempted to understand and explain the human condition and conditions of knowledge in our time. ... [A] recent shift ... within the humanities from deconstructive critiques of the metaphysics of presence to critical analyses of the politics of identity ... on the level of individuals and groups constructed around such categories as race, ethnicity, gender, sexuality, and their various intersections ... acknowledges the fact that identities, cultural or otherwise, are constructed and lived out not just in intrapsychic or metatheoretical realms but in the material space of differently authorized and institutionally vested powers.[13]

THE LOCALIZED SITUATION: MEXICAN AMERICANS IN SAN ANTONIO

The complex and shifting identities of Mexican-American Hispanics in central Texas involve the merger of heritages from original home territories and current immigration sites through more than 300 years of power shifts and the complex hybridization of cultural habits, values and preferences, including those connected with housing and ordinary environmental elements.[14]

In ground-breaking fieldwork, Jyothi Rao has shown empirically how important the personal modifications of institutional housing are to these residents' senses of individual and group well-being and identity.[15] Although numerous studies have been undertaken and published in specialized research journals, little sustained, major research has been published on specific ordinary topics in a way that has had an impact on public policy. As Rao noted:

> HUD conducted a survey in conjunction with National Hispanic Week [in 1992]. A selective survey was done of books on housing to determine the number of pages specifically discussing Hispanics, Chicanos, and Mexican-Americans. The study reported that in a sum total of 3,952 pages from various books, only 7 pages discussed Hispanics. Another survey of books on Hispanics [in 1985] found that in a total of 2,295 pages only 30 mentioned the housing conditions of Mexican-Americans.[16]

Rao's traditional methods of documentation and her analysis of several San Antonio low-income housing clusters showed that Mexican-American residents marked group identity and cultural values by means of many elements inside and outside their dwelling units, especially through symbols which provide links with tradition.[17] Concerning the features outside the dwellings, she found that the details and gestalt of statuary, altars, and miniature grottos (dedicated to Our Lady of Guadalupe and other, personally important saints); plants and flowers; and yards and fences not only distinguished Mexican-American residences, but also were explicitly understood by their makers as important in their personal and communal lives and ways of organization.[18] Mexican-American public housing residents in San Antonio and Austin also regarded yards and fences as important, and preferred that the front yard be more open than the back, but they did not agree on the desirability of one fence type or pattern of placement.

> A specific percentage for the preference of Mexican-Americans could not be arrived at [in regard to front-yard privacy]. Most of the 231 families preferred open front yards. Among those who wanted privacy, the majority were renters.
>
> 73.4% of the Mexican-Americans preferred a backyard which was closed off from the surrounding neighborhood. Of these, the majority preferred a fence to a solid wall.[19]

Other work in this area has supported these findings and approaches. Oscar Cadena, in his research on family compounds in San Luis Potosi, Mexico, found that the walls that traditionally distinguished family space from that of neighbors and social street life, as well as the walls inside dwellings that demarked zones of privacy and shared family life, all manifested common shared values and specific ethnic differences that shifted in modern housing designs. Whereas, "in traditional houses in Mexico blank walls usually face the public streets, in modern ones lawns are fenced or walled in and windows have grilles on the outside and drawn curtains inside" — a result corroborated by research in Garza Garcia's recent development outside Monterrey.[20]

D.J. Stamp, in a study of elderly residents, found that the same characteristics did not pertain in regard to the ethnically diverse Anglo population of Texas.[21] Melvin Williams, investigating body posture and behavior as a means to claim spatial territory in Black Philadelphia, found that such behavior was recognized and agreed to only within very specific, and differing, classes in a neighborhood.[22] Mary Murnane established that low-income Mexican-American and Black candidates for Habitat for Humanity dwellings in Austin, Texas, overwhelmingly preferred the traditional vernacular dwellings as found in their neighborhoods to the possibility of the new designs for cohousing.[23] To continue this line of investigation, it seemed appropriate to do further focal, local analysis and narrative construction, as Foucault has advised,[24] rather than attempt any ungrounded universalization of ethnicity constitution.

STRATEGIES FOR INQUIRY

As a result of shifting time and power relations, distinctions among diverse Anglo populations in substantially controlled institutional settings in central Texas have been almost entirely erased. In a parallel way, differences among "subordinated" ethnic groups within the vaguer category of "Hispanic" are part of a larger, continuing study of Texas populations, in regard to which marked differences exist between sites where different mixtures occur of Spanish speakers from Cuba, Puerto Rico, and Central and South American countries.[25] To remain usefully empirical and localized, I therefore decided to focus only on San Antonio, and only on two basic distinctions: (1) differences between the social practices and identities of Mexican-American Hispanics and those of African-American and Anglo populations; and (2), at greater length, differences between the practices and identities of Mexican-American Hispanic residents in public housing in San Antonio, and the practices of local Mexican-American officials in the San Antonio Housing Authority (SAHA) and the Hispanic and Anglo authorities at the Department of Housing and Urban Development (HUD) in Washington.

In addition, as noted earlier, whereas traditional studies of institutional housing had concentrated on the forms and materials of major symbolic and functional elements, my research focused on more everyday features. For example, very little attention had been given to yard systems and the various private and semi-private/semi-public passageways in standardized publicly supported dwelling environments among Mexican- or African-American groups (though some traditionally oriented work had been done on the issue in Anglo suburbs and middle-class residential areas).[26] My research aimed to begin to address this shortcoming. And in my empirical inquiry and analysis, I took as my twin compass points the fundamental importance of resident participation and a critical attitude in the tradition of Henri Lefebvre and Manuel Castells. As Castells has written:

> The vast majority of downgraded workers and new laborers share an excluded space that is highly fragmented, mainly in ethnic terms, building

> defensive communities that fight each other to win a greater share of services, and to preserve the territorial basis of their social networks, a major resource for low-income communities.[27]

To observe, visually and verbally document, and interpret the fencing patterns and semi-public passageways in three standardized public housing environments in San Antonio, I undertook fieldwork and analysis using hermeneutical phenomenology and interpretative anthropology.[28] Since it is obvious that both the housing authorities and residents have attempted to control passage and establish or maintain contested identities and differences, the task was to describe and compare the types, forms and materials of institutional and personal fencing systems and their patterns in relation to social observation and control. The specific goal was to describe and analyze the differences and similarities of fencing and passageway systems within each housing environment, between each housing environment and its surrounding neighborhood, and finally among the three sites and contexts.

Although hermeneutical phenomenology has significant similarities to ethnography and participant observation techniques,[29] it also encourages direct dialogue with the residents about their intentions and the meanings of historically developed physical and cultural features.[30] The phenomenological component involves the observation of the physical features and the behavior patterns of people using the built environment.[31] In a complementary fashion, the hermeneutic aspect seeks to interpret the ways in which residents build and change the environment in order to accomplish something important to them — and then attempts to understand how these accomplishments are contested and resolved (or not) in relation to the goals and desires of other political and ethnic groups involved in the same environment.[32]

In addition, I employed the critical spatial analysis of Henri Lefebvre to interpret the political tensions between the spatial practices of the Mexican-American residents undertaken as part of their lived experience in the city and the conceptually driven implementation of social demarcation, separation and control on the part of the professional planners and housing officials in the course of their representations of space. Lefebvre has distinguished the spatial practices that provide the realm for people's practical and sensory social lives as they go about their ordinary work and productive activity, from the conceptually focused operations of symbols through which specialists systematically and analytically create mental spaces which appear as representations of space. As Lefebvre has argued, the story of the modern era (and the source of tension between ordinary people and modernist professionals) has come down to a basic distinction: even though spatial practice provides the basis for secondary abstraction of conceptual space, the modern world has inverted this relation so that people's lives are more and more driven by abstractions rather than existential social needs. As a result, the built world proceeds less and less from shared social values and needs, and more and more from the symbolic systems of conceptual planning.[33]

My research follows Lefebvre's attempt to provide an adequate empirical study of physical and social environments by documenting and analyzing the struggle between the identity that the Housing Authority intended to implement through its fences, and that which the Hispanic residents, in social self-affirmation and in reaching out to other Hispanic community members produced for themselves.[34]

The live questions here concern how groups construct and use material features to constitute and legitimize identities, especially because until recently relatively few theories and field studies "have commented on how people manipulate material objects in power relations."[35] According to David Sibley: "There are implicit rules of inclusion and exclusion in a built form that contribute to the structuring of society and space in a way which some will find oppressive and others appealing."[36]

Precisely this sort of story happens in Mirasol Homes. Here the San Antonio Housing Authority (SAHA) and the Mexican-American residents and neighbors at first appeared to agree on the sense of place and identity. But later they diverged, as the official fence (aesthetically and operationally appealing to the authorities) was contemplated as a tool of separation and, in turn, abandoned as an idea in favor of population dispersal.

At the cultural-symbolic level, anthropology explains the ritual importance of establishing a proper boundary so that a house or settlement is ready for human habitation, and it tells of the ways in which cultural symbols provide links with tradition.[37] And given the increasingly sophisticated empirical studies of the material means and practices of social surveillance and control through investigations of prisons, schools, hospitals, factories, houses, gardens and landscapes, it is scarcely necessary to argue that the fence (as the wall) — one of the primal means of enclosing and excluding, of establishing and perpetuating boundaries — is a means of power over identity and place.

The empirical question is how, for specific locations and groups, the identities are resolved or contested. Sibley has written: "In the interaction of people and the built environment, it is a truism that space is contested, but relatively trivial conflicts can provide clues about power relations and the role of space in social control."[38] This is all the more true, Duncan and Ley have noted in relation to Cape Town, when the conflicts are not trivial, as in cases where "space not only contained the other [of the dominating group], but was also a label of status or stigma, . . . as when the varying destinations of displaced and resettled households became cultural signifiers," or where identity is threatened by actions that dissolve solidarity with neighbors and others of the same identity.[39] Housing projects notoriously belong in the category of sites where such ethnic or racial struggles of erasure versus manifestation of "Other" identity take place, and Mirasol Homes represents a modest site where ethnic identity was successfully, if only temporarily, constituted, before the almost invisible incident of its erasure. Hence, it provides a telling case of what traditional dwelling and settlement studies need better to understand.

Obviously, the historical social production of space and identity involves disparate conscious and unconscious efforts to compose and mark race, ethnicity, gender, class, and so on, undertaken simultaneously by many groups. According to Duncan and Ley: "Contemporary work, particularly in popular culture, sees society as constituted by a plurality of cultures, some dominant, some marginal."[40] Hence the agreement among critical theorists that what is needed is a modification of the tendency derived from the Enlightenment to seek universal and univocal features. Not only do many poststructuralists epistemologically disagree with the possibility of valid universal essences, but others point out that even if some general characteristics hold for limited periods of time, places or groups, the spaces of modernity have increasingly become both homogeneous and fragmented.[41] As a result, both radical and moderate theorists and researchers tend to investigate the issues of place and identity by encouraging a plurality of voices to come forward, allowing local stories to appear as the sites where individuals and groups speak for themselves or through their own representatives.[42] According to Sibley:

> It is necessary to examine the assumptions about the inclusion and exclusion which are implicit in the design of spaces and places. The simple questions we should be asking are: who are places for, whom do they exclude, and how are these prohibitions maintained in practice? Apart from examining legal systems and the practices of social control agencies, explanations of exclusion require an account of barriers, prohibitions and constraints on activities from the point of view of the excluded.[43]

To cite but one example, author and critic bell hooks provides a strong voice in this regard today. She speaks from her Black, womanly experience of places and about her and her acquaintances' emerging and suppressed identities. In "An Aesthetic of Blackness: Strange and Oppositional," she began:

> This is the story of a house. It has been lived in by many people. Our grandmother, Baba, made this house living space. She was certain that the way we lived was shaped by objects, the way we looked at them, the way they were placed around us. She was certain that we were shaped by space.[44]

In "Black Vernacular: Architecture as Cultural Practice," hooks further reflected on a youthful exercise in which she was encouraged to image a dwelling place of her dreams. She found,

> . . . precisely the kind of assignment that was meant to deflect attention from political realities, from the class, race, and gender differences that separated and divided us from one another. . . . This would have been a radically different assignment had we been encouraged to think critically about the actual spaces we inhabited, the neighborhoods and houses that were our world. Had we been given such an assignment, we would have learned to think about space politically, about who controls and shapes environments. . . . Doing this assignment, we might have come face to face with the politics of property, not only who owns and controls space but the relationship between power and cultural production.[45]

As Sibley has pointed out: "One part of the problem, then, is to identify forms of socio-spatial exclusion as they are experienced and articulated by the subject groups."[46]

Specifically, in the case of documenting and interpreting the fencing patterns and uses in several San Antonio public housing clusters, I engaged not only Mexican-American residents but African Americans and Housing Authority personnel concerning the political stresses and agendas among the residential players, and between them and the various officials. The result is not the never-to-be-obtained illusion of "completely objective fact," but an analysis and narrative that tells the local story of how identities are asserted and contested through building, using, and tearing down fences.

ETHNIC SPACE AND IDENTITY: THE VIEW OF THE AUTHORITY

The San Antonio Housing Authority (SAHA) has discovered fences. That is to say, they have discovered fences in general, and one kind of fence in particular, that catches their fancy. The reasons are not hard to discern. One could say that, objectively and aesthetically, the type of fences in question are handsome. They are elegantly designed, well crafted, and painted an attractive hue that evokes the patina of aged copper; and they are a fine complement to now-popular postmodern designs. That they are large and heavy might be overlooked if they were to appear against the scale of often exaggeratedly large postmodern institutional structures — departmental headquarters or office buildings, for example. However, an aesthetic and formal dissonance results when these large, heavy fences are used in residential projects, which — according to San Antonio's overall dwelling typology and the currently reigning governmental and environmental research ideology of low-rise public housing — are supposed to present a much more modest scale and a perceptual lightness befitting a domestic scene promising hope and freedom.

It seems that, in Jacques Derrida's terms, the meaning of the officially preferred fencing not only is ambiguous, but also genuinely "undecidable."[47] In the thought and discourses of defensible space, autonomous art work, postmodern aesthetics, and a government that "controls and cares," the fence not only is beautiful, but also it is liberating (e.g., from crime and fear) and good — thus progressive and desirable. However, in the thinking and discourses of self-determination, which advocate using nonstigmatizing elements in everyday life, and when seeking to nurture localized identities and stories, such fencing is hegemonic and bad — thus offensive.

In any case, as hooks and others have noted, the issue of the aesthetic of the fences scarcely embodies an autonomous, asocial judgment in any Kantian sense. The fences are there to carry out an intended social practice: to fence housing-project residents in and others out. And it is toward this end, in the name of safety, that the Housing Authority has developed a preferred mode of enclosure and separation, one that runs contrary to those local, ethnically preferred fencing patterns which can organize individual and shared spaces in an a socially congruent manner based on self-chosen relation and identity.

Such a choice of fencing type may be traced to fundamental ideological positions. On a federal level, HUD under its former Secretary Henry Cisneros was substantially influenced by Oscar Newman and his idea of "defensible space." This meant the agency shifted its policy toward fencing to view it as a research-legitimated way to control problems in public housing in an urban context. Local Hispanic Officials in San Antonio (predominantly appointed to office by previous Republican administrations) have concurred in this view. Meanwhile, as is discussed in detail below, an initial Request for Qualification (RFQ) and project to renovate Mirasol Homes under Secretary Cisneros, using proven resident participation models, was replaced, without explanation, with a plan to demolish the project and disperse its residents (if there are no residents, there is no need to figure out how they should participate in community redesign).[48] As the context for the present project, it is also important to note SAHA conceptually and pragmatically replaced strategies for user participation and the development of environmental features promoting neighborhood continuity with more formalized, centrally planned and executed separation by fencing.

Of course, long-standing controversies between strategies of control by physical separation and environmental integration are still being debated among researchers (e.g., between Newman on one side and Cooper-Marcus, Sarkissian, and Whyte on the other), hence, the importance of continuing empirical analysis. But in San Antonio the general conceptual shift to defensible space by way of separation achieved by fences can best be understood by comparing several housing projects and populations. In the process, a less naive view may emerge, one that offers less certainty about the goals and means that should — or might — be desired for a given housing community. Even if Louis Wirth is correct that "civilization can be judged at least to some extent by the minimum housing conditions which a society will tolerate for its members,"[49] criteria for determining appropriate housing conditions still need to be determined, and the actual means of improving housing need to be weighed in relation to politicized issues of self-identity and self-determination.

ETHNIC SPACE AND IDENTITY: AFRICAN AMERICANS IN EAST SAN ANTONIO

Even though the African American population of San Antonio is not large and is often overlooked because the city is seen in terms of largely Hispanic and Anglo images, and even though African Americans are not the focus of this research, the dominantly African American, east-side Spring View Apartments are an important part of San Antonio's urban and political milieu.[50] The residents of this housing unfortunately live in a physically desolate and dangerous environment. Not surprisingly, as traditional research would indicate, this situation correlates with the absence of features that might help bring about and sustain a sense of strong individual and group identity and participation in community-building.

Thus, there are almost no external marks of individualization of dwellings or family space and life in the areas outside dwelling units. No plants or art works or names can be found beside the doorways; no territorial markers are evident — no little fences or grass or pets; and no additional materials or forms have been used to alter individual spaces. Gang tags are the only real symbolization of identity.

As to fences, SAHA has recently begun to consider them to be a possible answer to the project's dangers on an urban-design scale. To the east of the units is a no-man's land crossed by the MKT Railway's tracks. The empty space is regularly the scene of fights and murders between the gangs that control the Spring View Apartments territory and gangs from the other side of the tracks. Although the issue seems to be one of civic order and spatial relations, and no one supposes a local "Berlin Wall" would solve the problem of gang violence, still a "big fence" seems to be a kind of comforting thought to the housing authorities. Thus, these African-American residents and their desperate place in the city's fabric may provide a base for the study of the Mexican-American scene. In particular, there is little self-constituted physical or social community at Spring View Apartments, and the idea of new fences appears as a kind of hopeless attempt on the part of the authorities to control offensive activities and create defensible space.

THREE TYPES OF SAN ANTONIO HOUSING AUTHORITY CLUSTERS

The difference between the East San Antonio African-American housing complex and the near west-side Mexican-American housing can be seen by a brief comparison. Three housing projects with a dominantly Mexican-American population cover the range of typical projects for this general ethnic group. The first, Villa De San Alfonso Apartments, is a modest set of 29 buildings in a residential area on the near west side of town. SAHA is proud of these 1990 postmodern buildings, designed for the elderly, and of design features such as window bars and balconies. The housing is essentially the same as current private-sector designs, done on one scale for residences and on a larger scale for institutional buildings. Because they have the look of "current development" happening all over the country, rather than being identified with any particular region or sub-population, these structures connote "what is now desired" and "as good as any" in mainstream, commercialized American culture.

The second complex, also on San Antonio's near west side, comprises a one-block area of new dwellings built on the site of dilapidated housing that was razed on the northern edge of Phase III of the very large Alazan-Apache Courts (1,096 units in 292 buildings). This new complex has served as a model project for the current HUD policy of providing housing as a way-station only (as distinct from long-term residency). A carefully selected group of people who are also enrolled in employment and support programs reside in this unusually expensive and well-designed postmodern complex. On one side of the complex is a famous local redevelopment project, Plaza Guadalupe, which has been featured over the last fifteen years as a successful attempt to build on the base of a strong parish and earlier community arts projects to rejuvenate an ethnic neighborhood, encourage civic pride, and provide for local business activity, the arts, and health-care delivery. Here, environmental design was seen as a key to reenergizing a community and encouraging investment by the private sector. On the other sides of the complex are residential areas and small businesses, often in houses.

Although it had still not become clear to me after several site visits whether SAHA was interested in keeping "dangerous" intruders out of the complex or keeping residents safely locked in — because neither seems to be an obvious reality — the officially favored fences certainly are prominent **(Figure 9.1)**. In terms of identity, the fences constitute the marking "new SAHA housing project," and separate the residents from their neighbors. Clearly, in designing and building the complex, SAHA wanted to maintain control over its image and the larger built environment of those who would live there. Thus, as with Spring View Apartments, there are no signs of personalized identity or family-group territory. The message seems to read: "Be glad for and respect the new housing. While you are here, keep it (and yourselves) just the way things are given and intended to be."

The third near west-side complex, Mirasol Homes, is much larger and older than either of the two new complexes just considered, more on the scale of Spring View Apartments. It has been the recipient of recent renovation funding that has changed its roofing, not only to prevent leakage, but also to follow the findings of research which suggests that to avoid stigmatization building

Figure 9.1. The official fence has a strong image. (Photo by author.)

forms and materials should echo those in the surrounding urban context. Thus, the change from earlier institutional flat roofs to pitched gabled roofs like those of houses belonging to Mexican Americans across the street seemed sensible and successful. Similarly, improved lighting within the site helped with safety.

As noted earlier, Mirasol Homes was also to be the scene of an additional major renovation project, in which I participated.[51] At Mirasol there was a strong sense of community identity among the predominantly Mexican-American residents. This was clearly evident in the heritage of self-produced murals from the 1970s that have not been gang-tagged, by the well-used communal facilities and by the fact that residents glared at my research team, challenged our presence as outsiders, and gave us warnings and directions as to how to behave, even when we were in the presence of on-site administrators — that is, official authorities. In short (and again as traditional housing research indicates), the residents exhibited a high positive sense of ownership and territorial propriety, explicitly demonstrating the understanding that I and my colleagues did not belong among the various nonresident architects, contractors and workers who have come in and out of the project during the last several years. Residents regularly acted positively on behalf of community territory and, thus, community safety.

There also was a strong sense of connectivity between the Mirasol project and the surrounding neighborhood. Both the residents and the designers of the original HUD participatory renovation project had expressed interest in this connectivity. As noted, the positive sense of internal and external community connection and affirmation were nurtured and expressed by a variety of personalized features: plantings, sacred and secular yard art, outdoor furniture, recreational equipment, and cooking implements, as well as by behavior patterns — and, of course, by perimeter and internal fencing that corresponded to the neighborhood vernacular.

DETAILS OF FENCING AND SEMI-PUBLIC PASSAGEWAYS IN MIRASOL HOMES AND THE SURROUNDING NEIGHBORHOOD

The nature of the fencing around the perimeter of Mirasol Homes is telling in regard to the relation of the complex and its residents to the surrounding community. The mainly Mexican-American residents of Mirasol — and the earlier SAHA administration that "decided on" and installed these fences — had affirmed the continuity of Mirasol with the neighborhood in many ways: by allowing for easy passage of children into and out of the complex, both at play and to and from school; by similar provision of rides to area public, private-nonprofit, and religious services via agency and church vans and vehicles; by locating city bus stops on the perimeter of the complex, giving neighbors and Mirasol residents a common ground; by scaling Mirasol buildings appropriately and by stepping them down to one story at the edges of the complex to merge with neighborhood buildings and street furniture; and by providing internal project streets which are basically continuous with the surrounding city grid (except for a problematic and complex twist along the southeast side).

Fences provided a key element in this overall strategy of integration/differentiation. Project fencing originally allowed ordinary ease of entry to the community services at Mirasol's center and facilitated a vernacular knitting-together with neighborhoods to Mirasol's north and northwest by echoing their fence and yard patterns. Mirasol fencing also allowed communication between Mirasol and surrounding areas by means of a passageway torn into the fence between the complex and Our Lady of the Lake University to the northeast (which would like to have more connection with Mirasol); by means of the openness of Mirasol to neighborhood residences to the southeast; and by the vernacular normality of the fence that separates the complex from the backyards of neighborhood residences to the south.[52] It is very important to note that the original fencing norm was set by the surrounding community in the form of standard chain-link fence, common throughout the city and neighborhood.[53] The fences around Mirasol Homes, then, demark territory but do not establish a separation. Quite the contrary, the fencing distinguishes Mirasol Homes in a manner that affirms its continuity with its surroundings and that is clearly permeable, allowing those who belong to the community by virtue of their Mexican-American identity to pass back and forth from the complex to the neighborhood.

The accomplishment and maintenance of the belonging-together of Mirasol and the neighborhood is clear from the behavior of Mirasol and neighborhood residents. For example, on the northwest side of the complex, where a part of the chain-link fence has been pulled down, the opening functions as a handy and well-used short-cut in and out **(Figure 9.2)**. Clearly, this is not a contested or no-man's land, but a vital and well-functioning passageway. As he sits outside his house next to it, a "mayor," to use William Whyte's term, amiably keeps his eye on all who come and go though this opening. He dispenses greetings and information, provides a sense of surveillance and proprietary interest on behalf of the community, and generally knows and cares about what is going on. Again, the criteria of traditional residential identity and behavior are substantiated and clearly function to produce a sense of community and safety. Unofficial pathways fan out from this fence, opening through the complex and back again.

Internally, the fences and patterns of use are more complex. Mirasol has a benign policy toward pets and personal modifications through landscaping and planting, outside decoration and activity, and to the construction of fences — all of which contribute to residents' involvement in their dwellings and social milieu. The housing authority does have regulations concerning the material, size and location of any fences built, but these generally fall within what residents report they would imagine or implement anyway. As noted earlier, the local norm within the San Antonio area for those at the lower and mid-income levels, and for businesses generally, is chain-link fence that comes to waist or chest height.

Within Mirasol there are three patterns of fences and movement through the community. The first consists of internal areas that have no fencing so that the yard spaces of each dwelling flow together. Of course, many of these areas have other territorial delimitations, established by pets on chains, plantings, or clotheslines; but others have no markings at all to personalize or "claim" them.

Figure 9.2. An unofficial opening — the "mayor" preferred not to be photographed. (Photo by author.)

The research team was not able to establish any consistent or prevalent account of how these spaces are used, or whether they have personal significance for residents. And it was not clear whether there is further meaning to these spaces, or whether some (or which) of them are basically indifferent or miscellaneous areas. (It also seemed that some of these areas, especially when running across the back of a row of units, are allowed to be open and are used as crossing spaces.)

A second internal pattern at Mirasol was found to be that of yard areas that are strongly identified with given units, which are delimited by the allowed and affordable chain-link fences put up by the residents **(Figure 9.3)**. Some of these areas are small and intensely used; others are as large as possible, sometimes appearing individually, and sometimes in clusters as residents apparently combine to set off their backyards from those of other neighbors and the complex at large. These larger scale environments naturally affect the flow of pedestrian traffic and play within Mirasol. In the areas in front of units, there are adequate paved walkways to carry residents without problem or confusion; but in the areas between the backs of the units, through-passage is entirely blocked in some places and diverted to the open, continuous yard space of those units where residents have not erected fences.

The third spatial pattern at Mirasol can be seen in the overall structure of open and closed space within the entire complex — that is, the possibilities and variations allowed by the unfenced and fenced-in yard space. It should also be noted that on the west and north sides of Mirasol there are public and play spaces that are

Figure 9.3. Units are differentiated by fences and ornamentation. (Photo by author.)

part of the scene of public passage. And, on the west end especially (and to a lesser extent on the north), there are some areas between the backsides of units large enough that they apparently have a "yard area" associated with the units on the two sides and a "surplus" in the middle that serves as an unpaved walkway for east-west movement. This is a predominantly east-west informal passageway because the complex's site plan has long blocks and open spaces running east-west with streets reinforcing that continuity, as do the clearly secondary north-south streets and north-south flow of pedestrian openings to play areas located midblock.

This open middle space through the length of the long blocks and accepted passages between the backs of dwellings is manifest and acknowledged in the fencing where the space running through the middle of the block (through what would be "backyards") meets the sidewalk and the road going around the block **(Figure 9.4)**. Here, where the back lots on the block might be expected to be closed off completely, preventing any through traffic across the middle of the block behind the dwelling units, the opposite occurs: a fence runs between the edge of a dwelling's back lot and the perimeter sidewalk, but stops where open passage though the back lots goes, and then picks up again to close off the edge of the opposite back lot and continuing perimeter sidewalk. (If this were a new suburban development, the passageway though the backs of the lots would be conceptualized and valued as a "community greenbelt" or a "hike and bike trail.")

What was especially interesting to observe at Mirasol was how subtly such built and behavioral phenomena were integrated into the residents' life worlds.

Figure 9.4. Consensual passageway through the backyards. (Photo by author.)

Thus, open spaces, fenced spaces, and unmarked middle passable spaces — all in nonregular arrangements — were understood and easily used every day by both inhabitants of Mirasol and residents of the surrounding community.

CONCLUDING THOUGHTS

A basic comparison of Mirasol Homes with the other Mexican-American publicly supported San Antonio complexes shows that the Mirasol residents have an ethnic community identity partially accomplished and sustained through individual and group modifications in their environment in a way that the African Americans in Spring View Apartments on San Antonio's east side do not. At the same time, the Mirasol residents produce an identity that is continuous with and in solidarity with surrounding Mexican-American neighbors that is not found, whether because it is suppressed or it is absent, in the SAHA-designed and -controlled Alazan-Apache Courts or Villa De San Alfonso Apartments.[54]

Indeed, as initially investigated by Rao, the fences and social uses of complex space (along with yard statuary and flowers and plants not studied here) do mark a Mexican-American identity that holds to Mexican traditions and on the other hand embraces American forms and materials. As such, it reflects a process through which the two aspects have been melded in a particular way in San Antonio for over 300 years so that they are inseparable, and in this particular historical hybrid mark the simultaneous and self-supporting occurrence of one version of Mexican-American identity, sense of place, and material-built environment.[55]

At the least, this research has shown how corresponding complexities of use and behavior mark insiders and intimates, in a manner that confirms the findings of traditional built environment and behavior research. Familiarity with these patterns of behavior (and acceptance by other residents) when crossing through the complex's irregular and unpredictable mix of informal and intimate spaces, semi-private and semi-public routes, and back-yard and perimeter fence barriers confirms the importance of "local recognition," with its surveillance factor — which Jane Jacobs once described as one of the ways neighborhood residents and regulars recognize and police those who don't live or belong there.[56]

I contend, however, that the findings go further. I believe they begin to establish how patterns of built things and human action themselves constitute a shared sense of place and belonging — that is, common identity and difference with others, for those who dwell by way of these things and actions. I do not believe I have found any primal group identity and material culture that traces back in continuity to ancient origins. There is no story here of the persistence and diffusion of material culture (say, from interior Mexico to San Antonio), so beloved by cultural geographers. Instead, what is evident today is part of the developing story of Mexican Americans in urban settings, where fusions of material and cultural traditions continue. Nor do I believe that this research has somehow established the existence of a nonhistorical or noncontextual ethnic identity for Mirasol Homes' Mexican Americans. Indeed, I have no sense of what personal or group identity or ethnicity would amount to outside of the historical and social contexts (horizons) that constitute shared and differentiated worlds of multiple, and usually contested and fragile, spatial environments and ways of living.[57]

Rather, toward the end of the twentieth century, as the poorer Mexican Americans of central Texas go about their urbanized lives amidst the modern, technological and bureaucratic structures surrounding them, they use fences that connote "this is what low- and moderate-income Mexican Americans in central Texas use around their dwellings." In the setting, the ordinary chain-link fences connote that they are not SAHA's official fences. Nor are they those of the wealthy and powerful individuals and institutions for whom they work (when they do work). Nor are they the self-conscious or camp uses of chain-link fencing, as found in recent avant-garde architectural designs. More directly, the fences at Mirasol are the fences the residents' choose to have. Of course, other groups use these fences too. So what? Situationally, the residents of Mirasol Homes affirm and maintain their Mexican-American identity, in part, by the inflection of their boundaries — articulating their own differences from others (e.g., from SAHA or the wealthy) and their belonging together with their Mexican American neighbors via the kinds and uses of fences that establish the continuity of Mirasol and its community context. Such a hybrid type of fence construction, as well as the hybrid identities it supports, seems to operate porously over a larger area. What else could they do? What more could they do? According to Castells:

> At the other end of the spectrum [from the upper tier], segmented local networks, often ethnically based, rely on their identity as the most valuable resource to defend their interests, and ultimately their being.[58]

Thus, although its importance should not be exaggerated, Mirasol's residents did manage to shape a small place in a way that helped articulate and nurture their uniqueness. The vernacular fences and behavioral-use patterns of residents and neighbors in Mirasol were a temporarily successful production of identity and place which was enacted through the construction of, and respect for, subtle but clearly discernible patterns of open space and fencing.

Unfortunately, SAHA not only has at its disposal its officially preferred fences by which it can control residents' environments and behavior, but also the power to disperse residents and to interrupt their communal relations with the neighboring community. SAHA has today exercised this authority, and is in the process of razing a substantial part of the Mirasol complex. Here, the successful self-constitution of individual and social ethnic identity and the self-determination of communal relations and life are precluded by the authority of political institutions that produce environments according to ideology and concepts, rather than by allowing inhabitants themselves to practice their preferred ways of living and building.

NOTES

1. O. Newman, *Defensible Space*, New York, Collier Books, 1973, pp.1-2.

2. N. Evenson, "The Hybrid Metropolis: Western Influences in India," in J.-P. Bourdier and N. AlSayyad (eds.), *Dwellings, Settlements and Tradition*, New York, University Press of America, 1989, pp.425-426.

3. Although the research literature is too large to survey fully here, some of the major works are J. Jacobs, *The Death and Life of Great American Cities*, New York, Vintage Books, 1961; E.T. Hall, *The Hidden Dimension*, Garden City, NY, Anchor Books, 1969; R. Sommer, *Personal Space*, Englewood Cliffs, NJ, Prentice Hall, 1969, and A. Mehrabian, *Public Places and Private Spaces*, New York, Basic Books, 1976.

4. P. Oliver, *Shelter, Sign, Symbol*, Woodstock, NY, The Overlook Press, 1977, pp.8,36. Relevant to my analysis of the fence as identity marker, where a group deliberately uses the fencing norm of the surrounding society in order to signal continuity with the latter (itself a hybrid of modest-income Texans and Americans), Oliver notes (p.8), "The eventual [Tonga] village, with its cluster of huts walled protectively by a fence of magical substance, was itself a symbol of the newly-born community. Its protective role was magical only — the fence was not substantial enough to be a protective barrier. In its traditional forms and instantly identifiable building details it was a sign to neighboring villages of their large unification within the Tonga lands. A sign too, of the existence of the village itself, signifying the presence of its occupants and the nature of their way of life."

5. C. Cooper-Marcus and W. Sarkissian, *Housing As If People Mattered*, Berkeley, University of California Press, 1986, and J.B. Rao, "Providing Mexican Americans in S.W. Texas with Homes — Not Houses," Master's thesis, University of Texas, Austin, 1992.

6. K. Davoli, "Women's Identities and Possessions in Temporary Dwellings," unpublished research report, University of Texas Community and Regional Planning Graduate Program, 1990; M. Francis, "Gardens in the Mind and in the Heart," in H. Hoogalem et al. (eds.), *Proceedings of the Tenth Biennial Conference of the International Association for the Study of People and Their Physical Surroundings*, Delft, Delft University Press, 1987; Newman, *Defensible Space*; and "Defensible Space: A New Physical Planning Tool for Urban Revitalization," paper delivered at the EDRA annual meeting, Boston, Spring, 1995, and published in *JAPA* 61, 1, pp.149-155.

7. M. Murnane, "Detached Desires: Austin Habitat for Humanity and the Creation of a Cohousing Community," Master's thesis, University of Texas, Austin, 1993; and Newman, *Defensible Space* and "Defensible Space."

8. D. Crow (ed.), *Geography and Identity: Exploring and Living the Geopolitics of Identity*, Washington, DC, Maisonneuve Press, 1996.

9. H. Lefebvre, *The Production of Space*, Oxford, Blackwell, 1991; J. Derrida, *Margins of Philosophy*, Chicago, University of Chicago Press, 1982; M. Foucault, *History of Sexuality, Vol.1: An Introduction*, New York, Pantheon, 1978; and *Discipline and Punish*, New York, Vintage, 1979; M. Keith and S. Pile, *Place and the Politics of Identity*, New York, Routledge, 1993.

10. See, for example, A.D. King, (ed.), *Representing the City*, New York, New York University Press, 1996; Bourdier and AlSayyad (eds.), *Dwellings, Settlements and Tradition*; S.M. Low and E. Chambers (eds.), *Housing, Culture, and Design*, Philadelphia, University of Pennsylvania Press, 1989; and A. Gupta and J. Ferguson (eds.), *Culture, Power, Place*, Durham, Duke University Press, 1997.

11. W.J.T. Mitchell (ed.), *Landscape and Power*, Chicago, University of Chicago Press, 1994, and S. Zukin, *Landscapes of Power*, Berkeley, University of California Press, 1991.

12. Keith and Pile, *Place and the Politics of Identity*, p.2.

13. A. Bammer (ed.), *Displacements: Cultural Identities in Question*, Bloomington, Indiana University Press, 1994, pp.xii,xiv,xvi.

14. This study focuses only on Mexican Americans in public housing. Also see note 25.

15. Rao, "Homes for Mexican Americans."

16. Ibid. The Studies referred to are John I. Gilderbloom, "Hispanic Housing in the U.S.: A Review of Recent Literature," Statement at Congressional Subcommittee on Housing and Community Development, 99th Congress, *Housing Needs of Hispanics*, Washington, DC, U.S. Government Printing Office, 1985; and HUD, *Comprehensive Housing Affordability Strategy 1991-1996 (CHAS)*, Texas, U.S. Department of Housing and Urban Development, Office of Community Planning and Development.

17. Rao, "Homes for Mexican Americans," p.130.

18. Ibid., p.58 and figures 10-12; pp.74,100,102 and figure 74; pp.105,120,156.

19. Ibid., p.179.

20. O. Cadena, "An Architectural Vocabulary for Houses in San Luis Potosi," Master's thesis, University of Texas, Austin, 1987, p.28; and the supporting recent work of Jevon Hull, "A Morphology of Housing Development in Garza Garcia, Mexico," Master's thesis, University of Texas, Austin, 1998.

21. D.J. Stamp, "Housing for Older Adults," Master's thesis, University of Texas, Austin, 1990.

22. M.D. Williams, *On The Street Where I Lived*, Philadelphia, Holt, Rinehart, and Winston, 1981.

23. Murnane, "Detached Desires."

24. T. O'Conner, "Foucault and the Transgression of Limits," in H. Silverman (ed.), *Philosophy and Non-Philosophy since Merleau-Ponty*, New York, Routledge, 1988.

25. I was neither expert enough, nor trusted enough, by the people I interviewed to discern the more subtle tensions among Mexican Americans who immigrated, whether recently or hundreds of years ago, from diverse parts of Mexico — nor between those from southern Mexico and Guatemala or Nicaragua, and so on. Obviously, understanding such distinctions and relations is important; but, just as clearly, it remains for someone else to undertake.

26. R. Mugerauer, "Midwestern Yards," *Places*, vol.2, no.2 (1985), pp.31-38; "A Phenomenology of Porches," in D. Seamon (ed.), *Dwelling, Seeing, Designing*, Albany, SUNY Press, 1992; and G. Dreicer (ed.), *Between Fences*, Cambridge, MIT Press, 1996.

27. M. Castells, *The Informational City*, Oxford, Blackwell, 1991, p.227.

28. I was hired by the San Antonio Housing Authority as lead planner to conduct field observations, interviews, and community meetings with the residents of Mirasol Housing in order to develop and implement a new Demonstration Master Plan for the revitalization of the existing 500 units and to add 300 additional units. The governing HUD Demonstration-Revitalization Program emphasized resident participation in design and integration of physical facilities and social resources with the surrounding community. I acted in this capacity as a consultant (tucked into the contracted Joint-Architect Partnership of Fisher, Heck, Imbimbo, Inc. and Durand Hollis Architects) from fall 1994 to spring 1995. At that time SAHA abruptly decided to abandon the project, and instead demolish a substantial portion of the complex, dispersing the residents and proceeding to build a small number of detached units throughout the neighborhood in partnership with the private sector. Because this decision was delivered without warning or explanation at the time I made my preliminary report, and because it was obviously in bad faith with regard to the promises of participation that had been made to the residents, I resigned from the project at this point. An accounting of the venture is found in "Community Planning for Housing Rejuvenation: Community Participation in the Mirasol Housing Project, San Antonio," a paper I delivered at the annual meeting of the Association of Collegiate Schools of Planning, 1994, Phoenix, which was subsequently published in the *Working Papers Series* of the Graduate School of Community and Regional Planning, University of Texas, Austin, 1996. The methodology is explained specifically in that paper and is generally elaborated on in several of my other publications: R. Mugerauer, *Interpretations on Behalf of Place*, Albany, SUNY Press, 1994; and *Environmental Interpretations*, Austin, University of Texas Press, 1996.

29. Compare especially to ethnomethodology, critical ethnography, interpretative anthropology, and radical empiricism, as in M. Jackson, *Radical Empiricism*, Bloomington, Indiana University Press, 1994; G. Psathas, *Phenomenology and Ethnomethodology*, New York, University Press of America, 1989; and M.F. Rogers, *Sociology, Ethnomethodology, and Experience*, Cambridge, Cambridge University Press, 1983.

30. P. Freire, *Pedagogy of the Oppressed*, New York, Seabury Press, 1968; and *Education for Critical Consciousness*, New York, Seabury Press, 1973.

31. A.A. Kawi, "The Oasis of Farafra in the Eyes of Its Inhabitants," in Bourdier and AlSayyad (eds.), *Dwellings, Settlements, and Tradition*, pp.357-78; and M. Richardson, "Being-in-the-Market versus Being-in-the-Plaza," *American Ethnologist*, 1982, pp.421-436.

32. Mugerauer, *Environmental Interpretations*.

33. This principally is carried out in Lefebvre's seminal *The Production of Space*.

34. See Lefebvre, *The Production of Space*, pp.3,38-40,42,362. I also would argue that the Mexican-American residents' fences generate representational space — Lefebvre's

third category of imaginative and symbolic variation. Although that task is beyond what is possible in this short chapter, it can readily be seen how the fencing patterns of the lived social spaces challenge the officially dominant spaces and are parallel to, or participate in, the dwellings imaginative importance for the users.

35. R.H. McGuire and R. Paynter, *The Archaeology of Inequality*, Oxford, Blackwell, 1991, pp.4,5.

36. D. Sibley, *Geographies of Exclusion*, New York, Routledge, 1995, pp.xi-xii.

37. See Rao, "Homes for Mexican Americans," p.130.

38. Sibley, *Geographies of Exclusion*, p.xiii.

39. J. Duncan and D. Ley (eds.), *Place/Culture/Representation*, New York, Routledge, 1993, p.6.

40. Ibid., p.11.

41. This paradox is taken up by Lefebvre and many others, such as Keith and Pile.

42. See, for example, Duncan and Ley, *Place/Culture/Representation*, p.8.

43. Sibley, *Geographies of Exclusion*, p.x.

44. b. hooks, *Yearning: Race, Gender, and Cultural Politics*, Boston, South End Press, 1990, p.103.

45. b. hooks, *Art on My Mind: Visual Politics*, New York, The New Press, 1995, pp.145-146.

46. Sibley, *Geographies of Exclusion*, p.x.

47. J. Derrida, *Spurs*, Chicago, University of Chicago Press, 1978.

48. See details in Mugerauer, "Community Planning for Housing Rejuvenation."

49. L. Wirth, "Housing as a Field of Sociological Research," *American Sociological Review*, 1987, p.139.

50. Not only is the phenomena beyond the scope of my intended investigation of ethnicity and fencing, but also, admittedly, the problems of Spring View Apartments, its African-American inhabitants, and their complexity are outside my expertise.

51. See Mugerauer, "Community Planning for Housing Rejuvenation."

52. On the south side of the complex, the fence is still the usual sort of chain link. But it appears higher and more "serious" or "hostile" in the small, block-long regularly flooded area where it separates the back sides of the complex's units from the neighbors' backyards.

53. Of course, this leads to the research issue of the origins and meaning of this type of chain-link fence for low- and low-middle-income groups and businesses in San Antonio, in the Southwest, and in the U.S. This vernacular topic sorely needs to be taken up.

54. The preferred heavy official fencing has made only one appearance at Mirasol — around a separately funded children's activity center toward the middle of the complex; it has not spread, much less become a norm.

55. See Rao, "Homes for Mexican Americans."

56. Jacobs, *The Death and Life of Great American Cities*.

57. Indeed, in the current intellectual, cultural and political climate, the next step may be more simply arguing how those who are "not settled" or "rooted" may yet have their own genuine mode of dwelling. See R. Mugerauer, "*Homo Viator* Stigmatized: Nomads', Migrant Workers', and Refugees' Modes of Dwelling," in S. Cairns and P. Goad (eds.), *Other Connections: Migrancy and the Limits of Architecture*, Melbourne, University of Melbourne, 1997.

58. Castells, *Informational City*, p.227.

EPILOGUE

"The Reverse Side of the World": Identity, Space, and Power

Ananya Roy

In 1997, Arundhati Roy attained meteoric fame for her first novel, *The God of Small Things*.[1] I was in Calcutta at that time, working on a research project, and it was clear that Roy's global acclaim had won her national attention: one of the countless and everyday ways in which the local is globalized, not the least through media networks. Over the next couple of years, people who had never heard of Kerala read about the town of Ayemenem and its residents — Baby Kochamma and "the twins" and Comrade Pillai. They marveled at the words of Roy, a petite, brown-skinned, shy-voiced woman, who had procured a $1 million advance from her American publishers and had won the Booker Prize (she had seemingly conquered both imperial America and colonizing Britain!). In my classes at U.C. Berkeley, eager students would ask if I were related to her — after all, we had similar sounding names, and spoke English with ease despite our obvious brown-skinnedness, although this remained unsaid. The unfamiliar had become quite familiar.

The reason I took a liking to *The God of Small Things* was not for its compelling storytelling or for its glimpses of childhood memories that so resonated with personal ones. Instead, I found the novel memorable for the distinctive idiom of Roy's voice: she writes not in English, but in the English of a postcolonial India. It is what might be called a "hybrid" vocabulary and imaginary. Her characters, and their fates, are hybrid. Take, for example, Sophie Mol (half English, half Indian) lying in a coffin in small-town Kerala with her "Made-in-England go-go bag," "the smell of England in her hair." Or Kochu Maria, the "vinegar-hearted cook," who spends her time watching *WWF Wrestling Mania* and *Santa Barbara*. And Ayemenem itself, a town whose history included both "Anglophile" Pappachi (his legacy — a chocolate box full of cufflinks) as well as the "Black Sahib," an Englishman gone native (his legacy — the History House turned later into a ritzy hotel that Roy calls the Heart of Darkness).

More important, Roy's system of representation is itself hybrid. Thus, Sophie Mol's face is "pale and wrinkled as a dhobi's thumb."[2] The vendor at the Abhilash Talkies is the "Orangedrink Lemondrink Man." Small fish appear in "the puddles that fill the PWD potholes on the highways."[3] These are intensely Indian-Keralite-Ayemenem experiences articulated in English, undeniably and inevitably in English. It is an idiom that signals a colonial legacy, and it is a crucial signifier of contemporary Indian middle-class identity.

What is this idiom? Salman Rushdie, in his introduction to *Mirrorwork*, a collection of English-language Indian writing, has noted that for some, this idiom "will never be more than a post-colonial anomaly, the bastard child of the Empire, sired on India by the departing British; its continuing use of the old colonial tongue a fatal flaw that renders it forever inauthentic."[4] But, as Rushdie has also emphasized, it would be wrong to dismiss this writing as colonial mimicry. Instead, writers like Roy have a "distinctively Indian" voice, going "well beyond the Englishness of the English language."[5] Dare I call this voice hybrid?

English was the "second language" in my postcolonial Calcutta school, but we spoke our "first language" for only 40 minutes out of a six-hour school day. Thus, my mother came to ask me on a regular basis: "What language do you dream in?" She was always a little sad at my immediate response, "English" — even though she herself taught the Queen's English to brown-skinned Calcuttans for much of her career. English — of course, I dreamt in English. It was as "natural" as the part of me that craved a cup of tea at four o'clock each afternoon.

Stuart Hall has written about the ritual of drinking tea as a signifying practice of the "English" identity, forged in the crucible of colonialism:

> Not a single tea plantation exists within the UK. This is the symbolization of English identity. I mean, what does anybody in the world know about an English person except that they can't get through the day without a cup of tea? Where does it come from? Ceylon — Sri Lanka, India. That is the outside history that is inside the history of the English. There is no English history without that other history.[6]

Well, as a citizen of postcolonial India, I, too, shared in this English ritual. Englishness had "seeped" into my life, "like tea from a teabag."[7] But this was also a modified and reworked Englishness. The tea I craved was with a good shot of milk and spices, brewed strong — similar to the product now being bottled and marketed in American markets as "chai" by boutique coffeehouses and Starbucks. And I drank it while wearing the stiff, starched white uniforms that were meant to represent British order in a seemingly disorderly independent India, except that the dresses were always limp and gray in the Calcutta heat, indelibly marked by our Indianness.

Is this the hybridity that has been theorized to be such an integral part of the postcolonial condition? Is there something about this hybridized experience that may shatter the simple dualism of "same" and "other," the neatness of being purely English or authentically Indian? Is there a cadence and imagery to voices like those of Roy and Rushdie that subvert established vocabularies and narratives?

These are, I think, quite important questions, made possible by the use of hybridity as an organizing concept. However, it is not my aim here to explore these questions. In this chapter I am interested in a somewhat different interrogation. What does it mean to use the term "hybrid"? Are there "other" imaginaries that present different intellectual — and inevitably political — possibilities? Overall, I do not intend to undermine the concept of hybridity, but I do hope to point to some of its inherent limitations and thus limited applicability. Also, I will argue that although most of the chapters in this book invoke the idea of hybridity in one way or another, they go well beyond the boundaries of the concept. From my perspective, this is a welcome transgression.

My own transgressions draw upon some of the ongoing debates in contemporary feminist thought regarding identity and space. But my choice of this tool kit is not meant to add a gendered dimension to the racially charged concept of hybridity. Instead, I argue that the ways in which the analytical category of gender has been conceptualized, and contested, in feminist debates can usefully inform discussions about hybridity.

Feminism itself is, of course, not a stable repository of concepts and ideas. In fact, it has been marked by bitter contentions rather than enduring certainties. However, I would like draw upon only one strand of this larger debate, one that can be designated poststructuralist feminism. Quite broadly, this is a feminism informed by postmodernism and postcolonialism.[8] This tradition, as I discuss, also has strong resonances with post-Marxist cultural studies.[9]

Specifically, there are three elements of poststructuralist feminism that I think are useful in an interrogation of hybridity. The first involves its rejection of biological essentialism and its attempt to theorize the social construction of subject-identities. Second, recent feminist debates about identity have brought a confrontation with the thorny question of difference — indeed, with multiple forms of difference. Poststructuralist feminism does not provide a single way of fixing difference and domination; instead, its usefulness is that it draws attention to *how* hierarchies of difference are constructed, legitimated and negotiated — and thus to processes of identity-formation and to the mechanisms of power.[10] The third way poststructuralist feminism is useful is that it is increasingly concerned with the spatiality of social ontologies. Not only does this involve a postcolonial mapping of global inequalities, but it also raises questions of interaction between space and identity,between politics and place. In the following sections, I discuss how these aspects of feminist debates can help rethink the concept of hybridity.

ESSENTIAL IDENTITIES

In his prologue to this book, Nezar AlSayyad outlines how and why the concept of hybridity is useful in thinking about identity in nonessentialist ways. Indeed, the current popularity of the term is in keeping with the broader postmodern rejection of a universal subject and its authorial claims. I do not repeat these discussions here, for their key points are summarized in the prologue and have been covered in many recent publications. Instead, I want to examine the issue of nonessentialism in light of some specific case studies, and later in relation to feminist debates.

The term "hybrid" implies a syncretic identity — a "third space" — a way of being that undermines standard dualisms. It has thus been used to deconstruct the seeming purity of races and ethnicities, to deauthenticate essentialist claims of identity. The case studies in this book do this effectively, and furthermore they quite usefully show that such forms of hybridization predate our current interest in them. There is a wonderful liveliness to the tainted sites thus surveyed, a jostling and elbowing for cultural hegemony that at first glance seems well in keeping with the ideas espoused by hybridity theorists.

However, there is a crucial difference between emphasizing that identities are fluid rather than fixed, syncretic rather than singular, and assuming that *all* forms of hybridity are inherently and naturally nonessential. Is hybridity always "something new and unrecognizable," always "a new area of negotiation of meaning and representation," always rife with subversive possibilities?[11] I think the case studies in this book show that hybridity, like all cultural practices, involves discursive struggles. They also indicate that these struggles can be as reactionary as they can be progressive, and, furthermore, that hybrid identities can be deployed as essentialized claims to truth and power. In other words, instead of assuming nonessentialism, it might be more useful to think of identity as a process, congealing in moments of fixity, in attempts to guarantee dominance in the face of social and cultural contingency.

Colonialism is a particularly good example of this "fixing" of identities. There is now a large body of research that shows how colonialism not only was about oppressive rule, but also about the management of cultural difference. What is further important for those interested in the built environment is that colonialism quite clearly involved the sociospatial management of difference. One rather obvious strategy involved in this process was spatial separation, manifested in such forms as the *cordon sanitaire* and dual cities. Equally interesting were a host of disciplinary strategies that can today be termed "hybrid." Indeed, Anthony King's analysis of the bungalows of British India and Gwendolyn Wright's explication of the Arabisances and "new medinas" of French Morocco point to the hybrid nature of colonial design and architecture.[12] But what does it mean to think of colonial cities as cultural hybrids? Did these spaces afford a glorious liminality within which to imagine postcolonialism? Perhaps. Take, for example, the French-preserved *casbah* of Algiers and the guerrilla war of independence that was waged on its rooftops.[13] This *casbah* does seem to resemble the celebrated "third space" — a hybrid site that challenged the colonial rules of engagement, an outcome inherent in the contradictions of colonial design itself.

Surely this is too utopian a reading of colonialism and its spatiality. Very simply, the emphasis on hybridity detracts from questions of power. As Gillian

Rose has asked, what then is the difference between colonial and third spaces? She has noted that the idea of hybridity provides a "disembodied" and "unbloodied" view of colonialism.[14] In a similar vein, the battle waged in the Algiers *casbah* cannot be seen only as an epistemic subversion from within — it was also bloody and protracted warfare. For all the ambiguities of colonialism, there were also the sureties of a quite stable cartography of power. And hybridized architecture often acted as a regulatory mechanism intended to reinforce this map. The seeming cultural hybridity of French design in Morocco was, as Paul Rabinow has pointed out, a social technology to orchestrate and control cultural diversity — what he has called "techno-cosmopolitanism."[15]

It is also perhaps a mistake to interpret the hybrid spatial forms of colonialism as an undermining of the dualisms of West and East, of colonizer and colonized. Tim Mitchell's work has shown how colonial technologies of power, often expressed spatially, legitimated and maintained rigid dualisms. The museumification of the Orient came to be an inextricable part of the discursive construction of an essentialized "Other." It was this difference — epitomized by the exotic, but rationalized, native — that made possible the West as exhibition, that gave identity to the techno-regulatory framework of nineteenth century modernity.[16] "The museum depended on the exhibition to achieve its significance and authenticity . . . [and] the exhibition depended on the museum which was what it aspired to escape."[17]

If the West produced the Orient through these essentialized dualisms, then so did these dualisms shape how the Orient imagined the West, and thus itself. In late nineteenth century Egypt, Khedive Ismail and his Minister of Public Works, Ali Mubarak — both possibly exemplars of hybridized (read: Westernized) colonial subjects — drew inspiration from the "exhibition," specifically from Haussmann's Paris, to reconstruct Cairo.[18] Here, as in so many cases, cultural hybridity was meant to guarantee an essentialized national identity, to fix a place for Egypt in a modern world. This hybridized response was no less pure, no less essentializing than Dostoevsky's rejection of the "exhibition," epitomized in his diatribe against the Crystal Palace: "I am afraid of this edifice just because it is of crystal and can never be destroyed, and because one could not stick one's tongue at it on the sly."[19] In both cases, identity (of ambitious Egyptians and envious Russians) was mapped within the dualistic West-Orient coordinates of a cartography of power.

In feminist scholarship, similar tensions have enlivened Inderpal Grewal's discussion of British colonialism in India. She has pointed to the gendered exoticization of the "Orient" in British discourses, epitomized by the idea of the "harem." It was the sense of the mysterious and seductive promiscuity of the harem that made possible the idea of a stable, ordered English home and nation. Grewal has gone further, emphasizing how Bengali nationalists also defined a sense of "home" in relation to a Westernized "Other." The linchpin of this alternative social order was the virtuous Bengali woman, idealized as *grihalakshmi*, or goddess of the home. She was to be the moral opposite of the British *memsahib*, safely tucked away in the *antapur*, or inner space, of the Bengali home. It was precisely this ideology of domesticity that made possible a nationalist imaginary of *bharatmata*, or motherland, a frail and gentle, feminized nation worth dying for. As Grewal has noted, this essentialized nationalism, the articulation of a pure Bengali (and Hindu) symbol, had its roots, interestingly enough, in the cultural hybridity of a hegemonic Bengali elite, well-versed in Victorian morality debates.[20] It was this hybridization that Dipesh Chakrabarty has aptly labeled "Bengali modern."[21]

Such forms of cultural struggle are not unique to the colonial condition. Much recent interest in hybridity has stemmed from a broader attempt to make sense of the identity politics of postcolonialism. Like colonialism, this politics has often involved the sociospatial management of difference. But, as Jane Jacobs has emphasized, these sociospatial struggles now play out not in distant colonies, but in the intimate "heart of empire."[22] Jacobs' study of Spitalfields in postimperial London has shown how the struggle over gentrification was simultaneously a struggle over meanings. Thus, each group — from the Heritage Trust to the Labour Party — quite predictably sought to fix and police a set of singular neighborhood interests, thereby staking claims to territory. Particularly pertinent for my discussion was the way the Bangladeshi elite sought to package and market what may be termed their hybrid identity as an authentic, essential, ethnic experience: "Banglatown." In her discussion of this phenomenon, Jacobs has noted the resonances with the commodified creation of Chinatowns in many parts of the world.[23]

Here then one can see an ethnic elite strategically essentializing its hybrid identity and producing a "Bengaliness" reworked to meet the imperatives of Thatcherite consumption. A similar dynamic can be seen in Katharyne Mitchell's study of the Hong Kong Chinese diaspora in Vancouver. Mitchell has shown how Chinese businessmen have presented themselves as "quintessential hybridized middlemen" — Chinese Canadian bridge-builders, "manipulating images of both the transnational, transcultural cosmopolitan and the ethnic Chinese, enabling them to position themselves on the margins of the nation, but at the lucrative center of Pacific Rim business."[24]

My intention in drawing upon these colonial and postcolonial examples is not to argue that difference, or the enactment thereof, is inevitably oppressive and reactionary. Instead, I wish to emphasize that the assumption of a liberatory hybridity is problematic, and that the question of subversion must remain just that — a question — rather than a pregiven conclusion.

Perhaps nowhere is this more apparent than at the current moment of globalization. For all the tainting of identities, for all the diasporas stretched across the globe, this moment is precisely the one at which pure, authentic, and even fundamentalist identities are being recovered.[25] Many of today's global battles are being localized in this essentialist idiom. A case in point is the string of food riots in southern France. Directed at the globalization of agriculture, these protests have targeted a very specific enemy, McDonald's, which has come to be seen as the symbol of U.S. imperialism and foul American-style food. The attempts of McDonald's to declare itself a local company — a hybrid, if you will — have gone unheeded, as French farmers have fought the globalization (read: Americanization) of taste. These cultural struggles have only been intensified with the recent imposition of stiff U.S. tariffs on such French food items as Roquefort cheese. French towns in the Roquefort-producing region have responded by imposing a symbolic 100 percent tax on Coca-Cola, another "American" company.[26]

Of course, there are interesting nuances to this story. McDonald's continues to thrive in France, even though the French refuse to admit to eating its quarter-pounders. As one French magazine claimed, "Spotted under the golden arches, the French often behave as if they've just got caught leaving an X-rated movie."[27] It is hard to ascribe nationality to the global megacorporations now dominating industrialized agriculture. Indeed, they are the true hybrids of the World Trade Organization (WTO) age, eluding such embodied and place-bound events as 1999's "Battle of Seattle." Nevertheless, there are culturally specific, essentialized symbols like McDonald's and Roquefort that mark this terrain of war. Thus, undergraduate students I have taught at U.C. Berkeley always smirk when they hear that José Bové, a sheep farmer leading the French protests, in fact has American connections (Bové's parents were once researchers at U.C. Berkeley, and he spent his early childhood in the Bay Area).[28] The students inevitably want to interpret this as a vindication of the "American" way of life rather than the fluidity of "French" identity.

I think that this, and the previous examples, point to the need to think about identity in open-ended and nonessential ways, of "hegemonic formations" as lived practices rather than guaranteed and stable structures.[29] The examples also indicate the imperative to take account of moments of articulation, the construction and legitimation of a pure and authentic "us," the congealed fixity that seems to defy negotiation.[30] This is precisely what poststructuralist feminism seems to offer, a way of theorizing identity and power in tandem.

At the core of current debates in feminism lies a rethinking of the notion of subject-identity, a rejection of the idea of stable and universal subjects. As Judith Butler has emphasized, this critique is "not a negation or repudiation of the subject but rather a way of interrogating its construction as a pregiven or foundationalist premise."[31] In particular, poststructuralist feminism has drawn attention to the ways in which the subject is "regulated and produced."[32] In other words, feminism has shifted the question from "What is the subject?" to "How is the subject constituted?" And it is this latter question, the issue of *how*, that the concept of hybridity seems unable to address. Hybridity takes as its starting point a pregiven subject. Despite all its ambivalences, it ultimately assumes that difference is a solved political problem, that the uneasiness of colliding systems of differences has been resolved in a neat and identifiable form of "new" identity, a specifiable subject-position. Poststructuralist feminism, on the other hand, is concerned with how the subject itself is the site of often irreconcilable differences, and how this means that identity is constantly reworked in open-ended and contingent ways.

I think that this feminist concern with difference is extremely important. The concept of hybridity seems to assume that there is a single axis of difference/power that can be subverted within a third space — a static syncreticism, if you will. But poststructuralist feminism shows how this singular mapping is limited, instead highlighting a complex and shifting geometry of difference and power. In each of the cases outlined earlier, for example, there exists multiple forms of difference, an intricate interweaving of class, gender and race hierarchies. Henrietta Moore has noted how contemporary feminist debates have been particularly concerned with these "multiple and contradictory positionings, subjectivities and discourses."[33] Not surprisingly, the title of her book is *A Passion for Difference*.

Feminist theories do not offer up an easy resolution to these issues. Instead, they insist that this heteronomy cannot be reduced to a single axis of difference/power. I would argue that it is precisely this dynamic intersectionality that the concept of hybridity has been unable to theorize or confront. For example, contrast the politics implicit in hybridity with Iris Young's rowdy and inexhaustible "politics of difference."[34] Stating "difference" clearly does make a difference. These feminist notions also resonate strongly with writings in post-Marxist cultural studies, which emphasize that "culture" is not a set of stable roles and identities, but rather a "constitutive social process," an inextricable part of which is the constitution and transformation of subject-positions.[35] Such ideas politicize the ontology assumed by hybridity, and thereby render the question of power and subversion open-ended.

What may be particularly useful about poststructuralist feminism in this discussion is that it situates hierarchies of difference within broad axes of social stratification. As Donna Haraway has noted, "Some differences are playful, others are poles of world-historical systems of domination. Epistemology is about knowing the difference."[36] Nowhere is this epistemological resolve more apparent than in the work of postcolonial feminists who have skillfully shown how subject-positions must be located within global systems of power.[37] In contrast to postmodernism's sense of undifferentiated otherness, these writers have stressed the profound unevenness of the global terrain, the sense that not all forms of difference are the same. Some of these ideas are hinted at in the concept of hybridity, which traces its roots to nonfeminist postcolonial theory, but none of them are explicitly theorized. In contrast to hybridity's utopian celebration of difference, postcolonial feminists like Gayatri Spivak have asked: "Can the subaltern speak?"[38] Such a question forces a confrontation with the conditions under which subject-positions are transformed into speaking-positions — as well as with the possibility of silence.

The case studies in this book seem to map a complex geometry of difference that is more in keeping with feminist cartographies than with hybrid ones. In doing so, they explore what Ernesto Laclau and Chantal Mouffe would call the "contingent logic of the social,"[39] a politics of difference whose outcomes are less certain than cultural hybridity. Let me briefly return to one of my cited examples to explain this point.

Katharyne Mitchell's study of the Hong Kong Chinese diaspora in Vancouver focused mainly on the complex identity struggles between "local" Canadians and Chinese businessmen, articulated in battles over urban space. But another set of struggles has intersected with this one. For example, the *New York Times* of August 29, 1999, reported how there was a growing fear in Vancouver of unchecked Chinese immigration, of boatloads of poor and illegal immigrants. "We are being treated like suckers!" said one interviewee. What makes this statement worthy of attention is that the speaker was Derick Y.J. Cheng, the owner of a Chinatown drugstore, chairman of Vancouver's Chinese Cultural Center, and a native of Fujian, the home province of many of the recent undocumented immigrants. Cheng's anger was echoed in an outpouring of letters to Vancouver's three Chinese dailies and on talk shows on local Chinese radio and television stations. The "business people" clearly had little sympathy for the diaspora of "boat people."[40] What can be made of the anger of this hybridized elite? What is the subject-position that allows them to speak so vociferously?

One way of thinking about this case is through Patricia Zavella's idea of an "insider-outsider" positionality.[41] Here it is hard to find a liberatory third space, with the promise of a stable resolution of difference. Instead, individuals and groups are simultaneously "inside" and "outside," uneasily implicated in multiple forms of difference, and the attendant exclusions and privileges. In the next section, I try to explore the spatiality of this seemingly paradoxical social ontology.

THE SPATIALITY OF HYBRIDITY

Is there a spatiality to hybridity? What kinds of spaces can be termed hybrid? Does hybridity have a unique geography? These are all provocative questions that Nezar AlSayyad addresses in his introduction. I would like to explore similar questions in light of feminist debates regarding identity, power and space, extending the feminist challenge to the core assumptions of hybridity. In so doing, I take up two interrogative themes that are central to contemporary feminist geography: How does the poststructuralist rethinking of subject-positions reconfigure ideas of space and place? And, conversely, how does spatiality figure in the politics of difference — that is, how is the social spatially constructed?

Perhaps the most obvious spatialized dimension of the concept of hybridity is the metaphorical invocation of a "third" space — the sense of a liberatory, in-between realm. There is the hint of a more concrete cartography here — the implication that hybridity involves a crossing of borders, some national — but this idea has remained quite abstract. For example, as AlSayyad discusses, the use of the concept of "borderlands" in recent hybridity debates has been largely metaphorical.

What kind of place is this third space? Katharyne Mitchell has rightly noted that the liminality of third spaces cannot automatically be assumed to be subversive; indeed these spaces can also be reactionary and oppressive. She has suggested that much more careful attention be paid to "the actual geographies of capital accumulation in which those spaces are produced."[42] At the heart of this call for a real geography — the urging to "ground" spatial metaphors in real places and times — is the intellectual imperative to take identity and power seriously.[43] But the idea of "third space" is problematic not only because it is a metaphorical invocation of space, but also because, as a metaphor, it assumes a homology between identity and space, conjuring up a utopian space that perfectly mirrors hybrid identities. In so doing it reifies both identity and space, bypassing the more messy geography of power/difference relations. Nowhere has this been more apparent than in the growing fascination with borders as the ultimate third space.

What are the borders that have thus captured the intellectual imagination? Borders, unlike boundaries, signify the copresence of identities foreign to one another, a confrontation with inalienable difference. But borders also hint at the possibilities of crossing over, of violating and redrawing limits. They carry with them, as Mitchell has commented, "the febrile fascination and flavor of the illicit."[44] This sense of sensuous transgression informs the opening quote of AlSayyad's prologue, the idea of the U.S.-Mexico border as a hybrid space, a frontal collision between cultures and identities with exciting opportunities for crossing over, being in-between. This specific border has been particularly subject to such forms of romanticization, as the following passage from *National Geographic* indicates:

> Whether or not either side likes it, America is being Mexicanized by immigrants (salsa now outsells catsup in the US) and Mexico is being Americanized by commercial competition and mass culture. The border is where you see this transition most dramatically. But even as the two countries become more alike, the border, *la frontera*, has become different from either. "Here, on the frontier, it's the frontier that counts," said an American bush pilot, never licensed, who spent his career flying for a Mexican rancher. "The rest of the country can go to the devil."[45]

This is a powerful and seductive idea, the sense of the border as the ultimate liminal zone, free of the restrictions of either country — where Scott, a San Diego judge, can lie in the arms of a Mexican prostitute, or where an unlicensed Texan pilot can soar through the sky. But to whom does this freedom extend? Surely not all are equally "free" at the border. Clearly this border is crossed with more ease by some than by all. If identity is a process, involving moments of essentialized fixity, then who gets to be hybrid? Who has the power to be hybrid? Who has the ability to articulate a hybrid identity as a social and political category, and when and under what conditions is this possible? For whom is the border a limitless frontier, and for whom is it an international boundary, heavily policed and regulated? To address these questions, one must take account of the complex geographies of the "real" border. Here, once again, I draw upon feminist insights, not only because of the gendered and sexualized nature of border narratives, but also because I believe that feminist debates provide the possibility of mapping power/difference relations.

Contrast the two hybrid border narratives — that of Judge Scott and the bush pilot — with the story of Rosario, a *maquiladora* worker, as told by Fernandez-Kelly.[46] Eighteen-year-old Rosario lives in Ciudad Juarez at the U.S.-Mexico border with her mother, a young brother, and two aunts in a two-room adobe house which leaks incessantly. All the women work in clothing *maquiladoras*, a grueling regimen of assembly-line production and measly wages. Rosario originally came to Juarez with dreams of the border in her head, enchanted with an architecture student called Carlos. On their first date, he had taken her to a European-style café and had ordered strawberries with cream. Rosario said that she had two servings, "just like an elegant lady in a romantic novel." She was pregnant at sixteen and shortly thereafter deserted by Carlos.

For Rosario, as for the San Diego judge and Texan daredevil pilot, the border holds possibilities of freedom — even sexualized freedom. Part of this possibility perhaps rests in a hybridized experience — of feeling like a European woman eating strawberries with cream, of seeing the twinkling lights of El Paso from the squalid shacks of Juarez, of buying the Avon cosmetics sold in the *maquiladora* neighborhoods. But Rosario's border is a far cry from a liberatory third space, and her choices at the border are as dreary as her daily routine of work. In this cartography of hegemony, the only freedom seems to rest with the *maquiladoras*, companies that have border-hopped with amazing ease, carving out a liminal space free from place-bound laws and regulations.

Are there other ways, then, of thinking about borders and boundaries, of imagining the spatiality of difference? Let me start with the issue of spatial metaphors, because metaphors indicate systems of representation and in turn political possibilities. Feminist geographers have provided spatial metaphors that contrast in interesting ways with the idea of third space. One such metaphor is Gillian Rose's concept of paradoxical space, by which she has sought to indicate the simultaneous occupation of mutually exclusive spaces: center and margin, inside and outside. Using this idea, Rose has tried to undermine the epistemological and ontological dualisms of what she has called masculinist geographies; and in this sense, her ideas have resembled notions of hybridity.[47]

Does it matter, then, which metaphor one chooses to use — third or paradoxical? Is there a difference between the two? Perhaps. The idea of third space seems to indicate a resolution of difference, a clear and fixed alternative free of contradictions — a borderland with its own identity. By contrast, Rose has used the idea of paradoxical space to indicate heteronomy, an implication in multiple, and often irreconcilable, hierarchies of difference. Her sense of being at once at center and margin, inside and outside, also resonates with Patricia Zavella's idea of insider-outsider positionality. Such feminist metaphors point to the ambiguous, contingent and contradictory privileges and exclusions of spatialized identities. This also seems to be what Chantal Mouffe has signaled in her concept of "contradictory interpellation": a provisionally fixed ensemble of subordinate and dominant subject-positions that indicates the multiplicity of social relations through which social agents are constituted.[48]

Zavella's and Mouffe's direct engagement with questions of identity and power also show how feminist theorists have gone well beyond metaphorical invocations of space. In this regard, I would like to briefly address two key feminist contributions I think are in keeping with the spirit of this book. The first has to do with the way in which places are conceptualized. The second is concerned with the spatiality of power/difference relations.

In the previous section, I outlined how poststructuralist feminists have emphasized the processual aspects of identity: the ways in which subjects construct, legitimate and rework multiple hierarchies of difference such as class, gender and race. Such ideas, Doreen Massey has argued, can be usefully extended to concepts of space and place: "It enables the problematization of any attempts at the construction of the identities of place as singular, bounded, static. And it opens up the possibility of talking about the presence of the global in the local."[49] Massey has emphasized the importance of paying attention to the "real multiplicities of space-time," to places as processes.[50] Her argument has thus shifted the focus from localities with inherent meaning and static identities to localizations — the ways in which localities are negotiated through the simultaneous constitution and contestation of subjects and communities.

Such ideas provide a useful way to return to the question of hybrid identities and spaces. If one thinks of cultural hybridity as an open-ended process, congealing in moments of essentialized fixity, then one can also think of hybrid places as fluid spaces with nodes of articulation, the spatio-temporal points at which the pure and authentic are distilled from amidst cultural corruptions. Thus, in Jacobs' example of Bengali-dominated Spitalfields, neo-Georgian "heritage houses" are preserved, and Bengaliness is managed and packaged in the form of a Banglatown. Likewise, in Mitchell's Vancouver, the "locals" designate "heritage trees," fertile symbols of a Canadian future, and Chinese businessmen eager to assert their rootedness reinscribe this imagery, presenting themselves as "trans/plants." And in southern France, McDonald's and Coca-Cola are defined as out of place, foot soldiers of an American invasion of French culture.

This is also where I think the case studies in this book excel. They show how the struggle for hegemony has involved the simultaneous and contested constitution of space and identity. The examples describe tainted sites, rich with the nuances of cultural hybridity, but their focus is repeatedly on power-laden attempts to recover authenticity, on how the essentialized hybrid is presented and legitimated as pure and genuine. That these attempts are often shaky only points to the enduring tension between the diasporas of people and ideas that hybridize places and the constant attempts to localize and fix the meanings and symbols of such spaces. In other words, the case studies have a messiness to them, a surplus, a contingent logic of the social, that cannot be contained in the idea of hybridity.

Herein lies the second point I wish to stress in relation to feminist debates. Feminist geographers have emphasized that this contingent logic of the social is inevitably negotiated in space and place. Thus, Massey has extended Mouffe's ideas of hegemony, showing how — along with identity and power — spatiality is a third interactive dimension.[51] In a similar vein, postcolonial feminists have situated postmodern discussions of difference in world-geographies and histories of domination. Such a shift from alterity to gendered subalternity has drawn attention to the complex and concrete core-periphery relations that mark the global arena.[52] This is also, I think, the enduring strength of the case studies in this book: they all show how the social is spatially constructed. If Rose has called for a firmer conceptualization of the "social" in hybridity debates, a more embodied sense of lived practices, then the case studies in this book provide what James Holston would call an "ethnographic conception of the social," locating cultural practices in place and time.[53]

What is the difference that geography makes? If subject-positions are also speaking-positions, then it is imperative that the cartography of hegemony that makes such speech and voice possible be mapped. If hybridity presents the utopian possibility of third spaces, then the case studies in this book go quite a few steps further: they reveal the tangible conditions under which the margins of space-geometries of power can become speaking-positions.

Much of this book is concerned with a particular set of spatialized cultural practices: urbanism. In this sense, it is in keeping with a renewed attention to cities as a "strategic arena" for the consolidation and negotiation of hegemony, a unique site of citizenship.[54] This linking of cities and citizenship has a long and varied lineage — from Henri Pirenne's idea of "islands of freedom," to Perry Anderson's celebration of ancient *civitas* — bodies of citizens.[55] In feminist writings, such as the work of Iris Young, cities have been recovered as sites of a politics of difference, places of "unassimilated otherness."[56] Unlike the third space idealized by the hybridity debates, Young's concept is marked by unresolved differences, the "being-together" of strangers, in an environment of aesthetic and social inexhaustibility. Here, difference is encountered and enacted in a "primal modernism" reminiscent of Baudelairean Paris.[57] Such inevitable and unsettling confrontations are present in many of the case studies in this book. And, as in the case of Young's politics of difference, the haunting presence of the Other is not easily mediated or transcended. Instead, it comes to signify the political possibilities of difference rather than the utopian dream of displaced liminality.

THE PRIVILEGE OF HYBRIDITY

All through this essay, and in keeping with contemporary feminist debates, I have tried to address three interwoven themes: identity, space and power. I have discussed the first two in relation to the concept of hybridity. In conclusion, I would like to explore the third by focusing on the discourse of hybridity and its attendant power relations.

In her review of Homi Bhabha's work, Gillian Rose has argued that Bhabha's convoluted textual style is the performance of a subject-position symptomatic of the contradictions of postcolonial discourse. In other words, Bhabha "writes the hybrid."[58] But what is this speaking-position? And what cartography of hegemony enables such speech? I think such questions explicitly raise the issue of knowledge/power relations, as well as define the imperative for self-reflexivity. Once again, I draw upon feminist debates to address these issues.

I started this essay with a discussion of Arundhati Roy's writings — what I designated as a hybrid idiom of representation because I think they indicate a particular speaking-position symptomatic of a specific historical moment. And I used the word "hybrid" because I believe that it is here, in the realm of discourse and representation, that the term has the most analytical edge. Very simply, I am arguing that the postcolonial possibility of a "third space" has borne fruit in an intellectual diaspora, seemingly free of the rigidities of place and locality, engaged in what Rushdie has called "a conversation with the world."[59] The "normalization" of hybridity that AlSayyad writes about in his prologue is perhaps related to this "postcolonial aura."[60] This is clearly an exciting set of possibilities, and one that should not be trivialized: the emergence of a particular set of subject-positions that deploy a new hybridized language and speak of a hybridized set of experiences. Trained in the Queen's English, I share in Rushdie's hope of "an English that goes well beyond Englishness."[61]

I emphasize this discursive aspect of cultural hybridity because I think it immediately points to a serious qualification that bears on the larger scope of the concept. It is, quite obviously, the question of power. Who has the power to be thus hybrid? To speak in a hybridized voice? To woo an audience thus? What is the world-geography of these speaking-positions? Such interrogations of privilege lead directly to the need for constant self-reflexivity. Feminist debates have a great deal to offer here. Adrienne Rich's idea of "a politics of location" is precisely about recognizing the geographic and historical coordinates of one's privileges — the privilege to be, to speak, to imagine.[62] Indeed, as Elspeth Probyn has noted, the very idea of speaking-positions raises the question of where and how one may speak.[63]

The issue of power is also quite nuanced. Can a voice such as Roy's simply be characterized as that of a postcolonial elite? Surely the stalwarts of India's Hindi heartland would like to dismiss her thus — as an "anglophile" Keralite. In many ways she resembles Zavella's "insider-outsider," a paradoxical ensemble of privileges and exclusions that belies the simplicity of hybridization. Feminists, in particular, have detailed the realities of this paradoxical intellectual space, the contingencies that mark the act of speech. I dare to speak within the institutional context of a First World academy, but with the unerased accent of the colonies that have learned to grow and drink tea. This is my borderland of sorts, locating me at once "inside and outside the project of academic privilege."[64]

However, as Pierre Bourdieu has insisted, there is also a broader imperative to reflexivity — one that forces recognition of the "intellectualist bias which entices us to construe the world as spectacle."[65] Perhaps it is this "bias" that lingers in the rather utopian idea of a "third space." Is it the dream of a free academy? Of the power of ideas? Of "consequent interventions"? Of a diaspora of philosopher-kings? I use the intellectualist constructs of feminism, a hybridity of sorts that allows me to suspend disciplinary rules of academic engagement. It is this hybridity that is perhaps the hardest to relinquish.

> A declaration that the world I am describing does have a reverse side, the possibility that I might find myself otherwise placed and orientated . . . a sign that the world presupposes a rest of the world. . . . I sense this reverse side of the world hidden beyond the deep thickness of earth and rock, and immediately it becomes the vertigo roaring in my ears and thrusting me towards the elsewhere.[66]

This declaration is a confrontation with the politics of difference, which perhaps in the arena of postcolonial intellectual diasporas is more unnerving than any on any city streets. It is a declaration to render the familiar unfamiliar, to admit to a reverse and different world, to admit to borders that cannot always be transgressed, and to an inalienable elsewhere that cannot always be tamed. In the spirit of this book, it is perhaps worth making note of Marshall Berman's reminder that this "reverse side of the world" is often within our own souls, a gnawing angst that once defined fin-de-siècle modernism and that seems to be present at this turn of the century as well.[67]

NOTES

1. A. Roy, *The God of Small Things*, New York, HarperPerennial, 1997.
2. *Dhobi* is the Hindi word for washerman.
3. PWD is an acronym for the Public Works Department.
4. S. Rushdie, "Introduction," in Rushdie and E. West (eds.), *Mirrorwork: 50 Years of Indian Writing, 1947-1997*, New York, Henry Holt & Co., 1997, p.x.
5. Ibid., p.xvi.
6. S. Hall, "Old and New Identities," in A. King (ed.), *Culture, Globalization, and the World-System*, Binghamton, SUNY Press, 1991, p.49.
7. Roy, *The God of Small Things*, p.33.
8. A sturdy discussion of these ideas can be found in N. Fraser and L. Nicholson, "Social Criticism without Philosophy: An Encounter between Feminism and Postmodernism," in Nicholson (ed.), *Feminism/Postmodernism*, New York, Routledge, 1990.
9. This encompasses a diverse group of scholars all concerned with Gramscian notions of hegemony. See, for example, E. Laclau and C. Mouffe, *Hegemony and Socialist Strategy: Towards a Radical Democratic Politics*, Verso, London, 1987; R. Williams, *Marxism and Literature*, New York, Oxford University Press, 1977; P. Willis, *Learning to Labor: How Working-Class Kids Get Working-Class Jobs*, New York, Columbia University Press, 1977; and S. Hall, "Cultural Studies and Its Theoretical Legacies," in L. Grossberg et al. (eds.), *Cultural Studies*, New York, Routledge, 1992.
10. J. Scott, *Gender and the Politics of History*, New York, Columbia University Press, 1988.
11. H. Bhabha, "The Third Space: Interview with Homi Bhabha," in J. Rutherford (ed.), *Identity: Community, Culture, and Difference*, London, Lawrence & Wishart, 1990, p.211.
12. A. King, *The Bungalow: The Production of a Global Culture*, New York, Oxford University Press, 1995; and G. Wright, *The Politics of Design in French Colonial Urbanism*, Chicago, University of Chicago Press, 1991.
13. M. Lamprakos, "Le Corbusier and Algiers: The Plan Obus as Colonial Urbanism," in N. AlSayyad (ed.), *Forms of Dominance: On the Architecture and Urbanism of the Colonial Enterprise*, Aldershot, Avebury, 1992.
14. G. Rose, "The Interstitial Perspective: A Review Essay on Homi Bhabha's The Location of Culture," *Environment and Planning D: Society and Space*, 13, 1995, pp.371–372.
15. P. Rabinow, "Colonialism, Modernity: The French in Morocco," in N. AlSayyad (ed.), *Forms of Dominance*, p.173.
16. T. Mitchell, *Colonising Egypt*, Cambridge, Cambridge University Press, 1988.

17. N. AlSayyad, "Contesting the Madina," *Islamic Culture*, vol.17, no.2, 1999, pp.1–17.

18. J. Abu-Lughod, *Cairo: 1001 Years of the City Victorious*, Princeton, Princeton University Press, 1971, p.105; and N. AlSayyad, "Ali Mubarak's Egypt: Between the Testimony of *Alamuddin* and the Imaginary of the *Khitat*," unpublished paper presented at the symposium "An 'Authentic' City for a Modern World: Cairo in the 19th Century," Center for Middle Eastern Studies, U.C. Berkeley, October, 1999.

19. F. Dostoevsky, *Notes from the Underground*, 1864, in M. Berman, *All That Is Solid Melts into Air: The Experience of Modernity*, New York, Penguin Books, 1982, p.236.

20. I. Grewal, *Home and Harem: Nation, Gender, Empire and the Cult of Travel*, Durham, NC, Duke University Press, 1996. See also P. Chatterjee, "The Nationalist Resolution of the Women's Question," in K. Sangari and S. Vaid (eds.), *Recasting Women: Essays in Indian Colonial History*, New Brunswick, Rutgers University Press, 1990.

21. D. Chakrabarty, "The Difference-Deferral of a Colonial Modernity: Public Debates on Domesticity in British Bengal," in D. Arnold and D. Hardiman (eds.), *Subaltern Studies VIII*, Delhi, Oxford University Press, 1994.

22. J.M. Jacobs, *Edge of Empire: Postcolonialism and the City*, New York, Routledge, 1996, p.38.

23. Ibid., p.99

24. K. Mitchell, "Different Diasporas and the Hype of Hybridity," *Environment and Planning D: Society and Space*, vol.15, 1997, p.551.

25. N. Smith, "Social Justice and the New American Urbanism: The Revanchist City," in A. Merrifield and E. Swyngedouw (eds.), *The Urbanization of Injustice*, New York, New York University Press, 1997, pp.122–123.

26. R. Cohen, "Heartburn: Fearful over the Future, Europe Seizes on Food," *New York Times*, August 29, 1999.

27. S. Daley, "French See a Hero in War on 'McDomination'," *New York Times*, October 12, 1999.

28. Ibid.

29. C. Mouffe, "Hegemony and New Political Subjects: Toward a New Concept of Democracy," in C. Nelson and L. Grossberg (eds.), *Marxism and the Interpretation of Culture*, Chicago, University of Illinois Press, 1988, pp.90–91.

30. C. Mouffe, "Post-Marxism: Democracy and Identity," *Environment and Planning D: Society and Space*, vol.13, 1995, p.263.

31. J. Butler, "Contingent Foundations: Feminism and the Question of Postmodernism," in Butler and J. Scott (eds.), *Feminists Theorize the Political*, New York, Routledge, 1992, p.9.

32. Ibid., p.13.

33. H. Moore, *A Passion for Difference*, Minneapolis, University of Minnesota Press, 1994, p.55.

34. I. Young, "The Ideal of Community and the Politics of Difference," in Nicholson (ed.), *Feminism/Postmodernism*.

35. Williams, *Marxism and Literature*, p.19.

36. D. Haraway, "A Manifesto for Cyborgs," in Nicholson (ed.), *Feminism/Postmodernism*, p.203.

37. C.T. Mohanty, "Under Western Eyes: Feminist Scholarship and Colonial Discourses," in Mohanty et al. (eds.), *Third World Women and the Politics of Feminism*, Bloomington, Indiana University Press, 1991.

38. G.C. Spivak, "Can the Subaltern Speak?" in Nelson and Grossberg (eds.), *Marxism and the Interpretation of Culture*.

39. Laclau and Mouffe, *Hegemony and Socialist Strategy*.

40. J. Brooke, "Vancouver Is Astir over Chinese Abuse of Immigration Law," *New York Times*, August 29, 1999.

41. P. Zavella, "Feminist Insider Dilemmas: Constructing Ethnic Identity with 'Chicana' Informants," *Frontiers*, vol.13, no.3, 1992, pp.53–76.

42. Mitchell, "Different Diasporas and the Hype of Hybridity," p.551.

43. N. Smith and C. Katz, "Grounding Metaphor: Towards a Spatialized Politics," in M. Keith and S. Pile (eds.), *Place and the Politics of Identity*, New York, Routledge, 1991.

44. K. Mitchell, "Transnational Discourse: Bringing Geography Back In," *Antipode*, vol.29, no.2, 1997, p.101.

45. R. Conniff, "Tex-Mex," *National Geographic*, February 1996, p.50.

46. M.P. Fernandez-Kelly, *For We Are Sold, I and My People*, Albany, NY, SUNY Press, 1983.

47. G. Rose, *Feminism and Geography: The Limits of Geographical Knowledge*, Minneapolis, University of Minnesota Press, 1993.

48. Mouffe, "Hegemony and New Political Subjects," p.95.

49. D. Massey, "Thinking Radical Democracy Spatially," *Environment and Planning D: Society and Space*, vol.13, p.284.

50. D. Massey, "Politics and Space/Time," *New Left Review*, vol.196, 1992, p.73.

51. Massey, "Thinking Radical Democracy Spatially," p.284.

52. Spivak, "Can the Subaltern Speak?"

53. G. Rose, "The Interstitial Perspective," p.373; J. Holston, "Spaces of Insurgent Citizenship," in L. Sandercock (ed.), *Making the Invisible Visible: A Multicultural Planning History*, Berkeley, University of California Press, 1998, p.55.

54. J. Holston and A. Appadurai, "Cities and Citizenship," *Public Culture*, vol.8, 1996, p.188.

55. H. Pirenne, *Medieval Cities: Their Origins and the Revival of Trade*, Princeton, Princeton University Press, 1925; and P. Anderson, *Lineages of the Absolutist State*, London, Verso, 1979.

56. I. Young, "The Ideal of Community and the Politics of Difference," pp.318–319.

57. Berman, *All That Is Solid Melts into Air*, p.148.

58. Rose, "The Interstitial Perspective," pp.365–367.

59. Rushdie, "Introduction," p.xiii.

60. A. Dirlik, "The Postcolonial Aura: Third World Criticism in the Age of Global Capitalism," *Critical Inquiry*, vol.20, no.2, pp.328–356.

61. Rushdie, "Introduction," p.xvi.

62. A. Rich, "Notes Towards a Politics of Location," in *Blood, Bread, and Poetry: Selected Prose, 1979–1985*, New York, W.W. Norton, 1986.

63. E. Probyn, "Travels in the Postmodern: Making Sense of the Local," in Nicholson (ed.), *Feminism/ Postmodernism*, p.177.

64. L. McDowell, "Speaking from Inside and Outside the Project," *Antipode*, vol.24, no.1, 1992, p.68.

65. P. Bourdieu, in Bourdieu and L. Wacquant, *An Invitation to Reflexive Sociology*, Chicago, University of Chicago Press, 1992, p.36.

66. I. Calvino, *The Road to San Giovanni* (English edition), New York, Vintage International, 1993, pp.146–147.

67. Berman, *All That is Solid Melts into Air*, p.153.

SELECTED BIBLIOGRAPHY

Abu-Lughod, Janet, *Cairo: 1001 Years of the City Victorious*, Princeton, Princeton University Press, 1971.

Abu-Lughod, Janet L., *Before European Hegemony: The World System, A.D. 1250-1350*, New York, Oxford University Press, 1989.

AlSayyad, Nezar, (ed.), *Forms of Dominance: On the Architecture and Urbanism of the Colonial Enterprise*, Aldershot, Avebury, 1992.

Anderson, Benedict, *Imagined Communities*, London, Verso, 1983.

Appadurai, Arjun, *Modernity at Large: Cultural Dimensions of Globalization*, Minneapolis, University of Minnesota Press, 1996.

Bammer, Angelika (ed.), *Displacements: Cultural Identities in Question*, Bloomington, Indiana University Press, 1994.

Bhabha, Homi K., *The Location of Culture*, London and New York, Routledge, 1994.

Biger, G., *An Empire in the Holy Land: Historical Geography of the British Administration of Palestine 1917-1929*, New York, St. Martin's Press, 1994.

Bowen, H. and H. Gibb, *Islamic Society and the West*, London, Oxford University Press, 1960, 2 vols.

Castells, Manuel, *The Informational City*, Oxford, Blackwell, 1991.

Castells, Manuel, *The Power of Identity*, Malden, MA, Blackwell, 1997.

Crow, Dennis (ed.), *Geography and Identity: Exploring and Living the Geopolitics of Identity*, Washington, DC, Maisonneuve Press, 1996.

Dirlik, Arif, *After the Revolution*, Hanover and London, Wesleyan University Press, 1994.

Featherstone, Mike, *Global Culture: Nationalism, Globalization, and Modernity: A Theory, Culture & Society Special Issue*, London, Sage Publications, 1990.

Featherstone, Mike et al. (eds.), *Global Modernities*, London, Thousand Oaks, CA, and New Delhi, Sage Publications, 1995.

Fludernik, Monika (ed.), *Hybridity and Postcolonialism: Twentieth Century Indian Literature*, Tubingen, Stauffenburg Verl, 1998.

Giroux, Henry, *Border Crossings: Cultural Workers and the Politics of Education*, New York and London, Routledge, 1992.

Grewal, Inderpal, *Home and Harem: Nation, Gender, Empire and the Cultures of Travel*, Durham, NC, Duke University Press, 1996.

Hall, Stuart and Paul Du Gay, *Questions of Cultural Identity*, London, Sage, 1996.

Hannerz, Ulf, *Cultural Complexity: Studies in the Social Organization of Meaning*, New York, Columbia University Press, 1992.

Harvey, David, *The Condition of Postmodernity: An Enquiry into the Origins of Culture Change*, Cambridge, Basil Blackwell, 1989.

Jacobs, Jane M., *Edge of Empire: Postcolonialism and the City*, New York, Routledge, 1996.

King, Anthony, *The Bungalow: The Production of a Global Culture*, New York, Oxford University Press, 1995.

King, Anthony D., *Colonial Urban Development: Culture, Social Power, and Environment*, London and Boston, Routledge & Paul, 1976.

King, Anthony D., *Culture, Globalization, the World System*, London, McMillan, 1991.

King, Anthony D., *Global Cities: Post-Imperialism and the Internationalization of London*, London and New York, Routledge, 1990.

King, Anthony D. (ed.), *Representing the City*, New York, New York University Press, 1996.

Laclau, Ernesto and Chantal Mouffe, *Hegemony and Socialist Strategy: Towards a Radical Democratic Politics*, Verso, London, 1987.

Lefebvre, Henri, *The Production of Space*, Oxford, Blackwell, 1991.

Massey, Doreen B., *Space, Place, and Gender*, Minneapolis, University of Minnesota Press, 1994.

Mitchell, Timothy, *Colonising Egypt*, Cambridge, Cambridge University Press, 1988.

Mitchell, W.J.T. (ed.), *Landscape and Power*, Chicago, University of Chicago Press, 1994.

Moore, Henrietta, *A Passion for Difference*, Minneapolis, University of Minnesota Press, 1994.

Mugerauer, Robert, *Environmental Interpretations*, Austin, University of Texas Press, 1996.

Oliver, Paul, *Shelter, Sign, Symbol*, Woodstock, NY, The Overlook Press, 1977.

Pred, Allan R. and Michael Watts, *Reworking Modernity: Capitalisms and Symbolic Discontent*, New Brunswick, NJ, Rutgers University Press, 1992.

Rutherford, J. (ed.), *Identity: Community, Culture, and Difference*, London, Lawrence and Wishart, 1990.

Said, Edward W., *Orientalism*, New York, Pantheon Books, 1978.

Werbner, Pnina and Tariq Modood (eds.), *Debating Cultural Hybridity: Multi-Cultural Identities and the Politics of Anti-Racism*, London and Atlantic Highlands, NJ, Zed Books, 1997.

Williams, Raymond, *Keywords*, New York, Oxford University Press, 1976.

Wright, Gwendolyn, *The Politics of Design in French Colonial Urbanism*, Chicago, University of Chicago Press, 1991.

Zukin, Sharon, *Landscapes of Power*, Berkeley, University of California Press, 1991.

INDEX

ABOUT THE CONTRIBUTORS

NEZAR ALSAYYAD is an architect, planner, urban historian and a professor of architecture and planning at the University of California at Berkeley. He also currently serves as chair of the Center for Middle Eastern Studies and director of the International Association for the Study of Traditional Environments at the University of California at Berkeley.

AYFER BARTU is an assistant professor of anthropology at Koç University in Istanbul, Turkey. Her areas of interest include anthropology of tourism, urban anthropology, heritage studies and social theory. She has published several articles on the politics of historic preservation in Istanbul.

ANNE-MARIE BROUDEHOUX is an architect who is a graduate of the School of Architecture at McGill University in Montreal, Canada. She is currently a Ph.D. candidate in architecture in the area of environmental design in developing countries at the University of California at Berkeley.

GREG CASTILLO is an assistant professor of architectural history at the University of Miami, Florida. He is currently working on a history of the transmission of American and Soviet design paradigms to postwar Germany, and has recently completed a Ph.D. dissertation entitled "Constructing the Cold War: Architecture and the Cultural Division of Germany" at the University of California at Berkeley.

RON FUCHS studied architecture at the Faculty of Architecture and Town Planning, Technion — Israel Institute of Technology, Haifa, where he received his M.Arch. and D.Sc. degrees. He has been lecturing on the history of architecture at the Technion since 1992.

THOMAS R. GENSHEIMER is an assistant professor in the Department of Architectural History, Savannah College of Art and Design, Savannah, GA. He recently received his Ph.D. in architectural history from the University of California at Berkeley, and has conducted research in Kenya on the urban history of East Africa as a Fulbright Scholar.

GILBERT HERBERT is a professor emeritus of architecture at the Technion-Israel Institute of Technology, Haifa. He has also held positions at the University of Witwatersrand in South Africa and the University of Adelaide in Australia.

ALI A. MAZRUI is the director of the Institute of Global Cultural Studies and Albert Schweitzer Professor in the Humanities at Binghamton University (State University of New York at Binghamton). He is the author of many books, articles, and television programs on Africa.

ROBERT MUGERAUER is Sid Richardson Centennial Professor of Architecture and Planning and also a faculty member in the Departments of Geography, Philosophy, and American Civilization at the University of Texas, Austin. His most recent books are *Interpretations on Behalf of Place* (1994) and *Environmental Interpretations* (1996).

ANANYA ROY was the Executive Coordinator of the International Association for the Study of Traditional Environments (IASTE) from 1994 to 1999. She recently received a Ph.D. in urban planning from the University of California, Berkeley, where she currently is a visiting lecturer in the Departments of City and Regional Planning and Women's Studies.

CHRISTOPHER L. YIP has done research on various aspects of Asian-American settlement and Asian architecture. He has been a research fellow at the Centre of Asian Studies, University of Hong Kong, and at the Institute for the Study of Social Change, University of California, Berkeley, where he also received a Ph.D. in architectural history. He is currently a professor of architecture at the California Polytechnic State University, San Luis Obispo.

SIBEL ZANDI-SAYEK is a Ph.D. candidate in the Department of Architecture at the University of California at Berkeley. She is working on a dissertation entitled "The Remaking of an Ethnic Landscape: Politics of Space and Identity in Nineteenth Century Izmir."